Biologically inspired textiles

The Textile Institute and Woodhead Publishing

The Textile Institute is a unique organisation in textiles, clothing and footwear. Incorporated in England by a Royal Charter granted in 1925, the Institute has individual and corporate members in over 90 countries. The aim of the Institute is to facilitate learning, recognise achievement, reward excellence and disseminate information within the global textiles, clothing and footwear industries.

Historically, The Textile Institute has published books of interest to its members and the textile industry. To maintain this policy, the Institute has entered into partnership with Woodhead Publishing Limited to ensure that Institute members and the textile industry continue to have access to high-calibre titles on textile science and technology.

Most Woodhead titles on textiles are now published in collaboration with The Textile Institute. Through this arrangement, the Institute provides an Editorial Board which advises Woodhead on appropriate titles for future publication and suggests possible editors and authors for these books. Each book published under this arrangement carries the Institute's logo.

Woodhead books published in collaboration with The Textile Institute are offered to Textile Institute members at a substantial discount. These books, together with those published by The Textile Institute that are still in print, are offered on the Woodhead web site at: www.woodheadpublishing.com. Textile Institute books still in print are also available directly from the Institute's web site at: www.textileinstitutebooks.com

A list of Woodhead books on textiles science and technology, most of which have been published in collaboration with The Textile Institute, can be found at the end of the contents pages.

Woodhead Publishing in Textiles: Number 77

Biologically inspired textiles

Edited by
A. Abbott and M. Ellison

The Textile Institute

CRC Press
Boca Raton Boston New York Washington, DC

WOODHEAD PUBLISHING LIMITED
Cambridge, England

Published by Woodhead Publishing Limited in association with The Textile Institute
Woodhead Publishing Limited, Abington Hall, Granta Park,
Great Abington, Cambridge CB21 6AH, England
www.woodheadpublishing.com

Published in North America by CRC Press LLC, 6000 Broken Sound Parkway, NW,
Suite 300, Boca Raton, FL 33487, USA

First published 2008, Woodhead Publishing Limited and CRC Press LLC

British Library Cataloguing in Publication Data
A catalogue record for this book is available from the British Library.

Library of Congress Cataloging in Publication Data
A catalog record for this book is available from the Library of Congress.

Woodhead Publishing ISBN 978-1-84569-247-6 (book)
Woodhead Publishing ISBN 978-1-84569-508-8 (e-book)
CRC Press ISBN 978-1-4200-7985-2
CRC Press order number WP7985

The publishers' policy is to use permanent paper from mills that operate a sustainable forestry policy, and which has been manufactured from pulp which is processed using acid-free and elementary chlorine-free practices. Furthermore, the publishers ensure that the text paper and cover board used have met acceptable environmental accreditation standards.

Typeset by Ann Buchan (Typesetters), Middlesex
Printed by TJ International Limited, Padstow, Cornwall, England

Contents

Contributor contact details

(* = main contact)

Editors

Professor Albert Abbott
Department of Biological Sciences
Clemson University
Clemson
South Carolina 29634-0907
USA

Email: aalbert@clemson.edu

Professor Michael Ellison
School of Materials Science and Engineering
Clemson University
Clemson
South Carolina 29634-0971
USA

Email: ellisom@clemson.edu

Chapter 1

Dr F. Teulé* and R. V. Lewis
Department of Molecular Biology
University of Wyoming
Laramie, Wyoming 82071
USA
Email: fteule@uwyo.edu

W. R. Marcotte Jr and Professor A. G. Abbott
Department of Genetics and Biochemistry
Jordan Hall, Room 116
Clemson University
Clemson, South Carolina 29634
USA

Email: aalbert@clemson.edu

Chapter 2

Associate Professor S. W. Harcum
Department of Bioengineering
Clemson University
Clemson, South Carolina 29634-0905
USA

Email: harcum@clemson.edu

Chapter 3

Dr F. Teulé
Department of Molecular Biology
University of Wyoming
Laramie, Wyoming 82071
USA

Email: fteule@uwyo.edu

Chapter 4

C. Holland* and Professor F. Vollrath
Department of Zoology
University of Oxford
Oxford OX1 3PS
UK

Email: christopher.holland@zoo.ox.ac.uk; fritz.vollrath@zoo.ox. ac.uk

Chapter 5

Dr C. Santulli
University of Reading
Centre for Biomimetics
Reading RG6 6AY
UK

Presently at:
Università di Roma
Dipartimento di Ingegneria Elettrica
via Eudossiana 18
00184 Roma
Italy

Email: carlo.santulli@uniroma1.it

Chapter 6

V. Kapsali and P. Dunamore
London College of Fashion
20 John Princes Street
London W1G 0BJ
UK

Email: v.kapsali@fashion.arts.ac.uk

Chapter 7

Dr-Ing. T. Stegmaier*, Dr V. von Arnim, A. Scherrieble and Professor Dr-Ing H. Planck
Institute of Textile Technology and Process Engineering, Denkendorf (ITV)
Koerschtalstrasse 26
D-73770 Denkendorf
Germany

Email: thomas.stegmaier@itv-denken dorf.de

Chapter 8

Professor L. Hes
Technical University of Liberec
Studentská 2
461 17 Liberec 1
Czech Republic

Email: luboshes@hotmail.com

Chapter 9

Dr M. Milwich* and Professor Dr-Ing H. Planck
Institute of Textile Technology and Process Engineering, Denkendorf (ITV)
Koerschtalstrasse 26
D-73770 Denkendorf
Germany

Email: Markus.Milwich@itv-denken dorf.de

T. Speck and O. Speck
Plant Biomechanics Group
Botanischer Garten
Universitaet Freiburg
Schaenzlestrasse 1
D-79104 Freiburg
Germany

Chapter 10

Dr-Ing T. Stegmaier*, Dr V. von Arnim, M. Linke, Dr M. Milwich, Dr J. Sarsour, A. Scherrieble, P. Schneider, Professor Dr-Ing H. Planck
Institute of Textile Technology and Process Engineering Denkendorf (ITV)
Koerschtalstrasse 26
D-73770 Denkendorf
Germany

Email: thomas.stegmaier@itv-denken dorf.de

Woodhead Publishing in Textiles

1 **Watson's textile design and colour Seventh edition**
Edited by Z. Grosicki

2 **Watson's advanced textile design**
Edited by Z. Grosicki

3 **Weaving Second edition**
P. R. Lord and M. H. Mohamed

4 **Handbook of textile fibres Vol 1: Natural fibres**
J. Gordon Cook

5 **Handbook of textile fibres Vol 2: Man-made fibres**
J. Gordon Cook

6 **Recycling textile and plastic waste**
Edited by A. R. Horrocks

7 **New fibers Second edition**
T. Hongu and G. O. Phillips

8 **Atlas of fibre fracture and damage to textiles Second edition**
J. W. S. Hearle, B. Lomas and W. D. Cooke

9 **Ecotextile '98**
Edited by A. R. Horrocks

10 **Physical testing of textiles**
B. P. Saville

11 **Geometric symmetry in patterns and tilings**
C. E. Horne

12 Handbook of technical textiles
Edited by A. R. Horrocks and S. C. Anand

13 Textiles in automotive engineering
W. Fung and J. M. Hardcastle

14 Handbook of textile design
J. Wilson

15 High-performance fibres
Edited by J. W. S. Hearle

16 Knitting technology Third edition
D. J. Spencer

17 Medical textiles
Edited by S. C. Anand

18 Regenerated cellulose fibres
Edited by C. Woodings

19 Silk, mohair, cashmere and other luxury fibres
Edited by R. R. Franck

20 Smart fibres, fabrics and clothing
Edited by X. M. Tao

21 Yarn texturing technology
J. W. S. Hearle, L. Hollick and D. K. Wilson

22 Encyclopedia of textile finishing
H-K. Rouette

23 Coated and laminated textiles
W. Fung

24 Fancy yarns
R. H. Gong and R. M. Wright

25 Wool: Science and technology
Edited by W. S. Simpson and G. Crawshaw

26 Dictionary of textile finishing
H-K. Rouette

72 Advances in wool
Edited by N. A. G. Johnson and I. Russell (forthcoming)

73 Military textiles
Edited by E. Wilusz

74 3D fibrous assemblies: Properties, applications and modelling of three-dimensional textile structures
J. Hu

75 Medical textiles 2007
Edited by J. Kennedy, A. Anand, M. Miraftab and S. Rajendran (forthcoming)

76 Fabric testing
Edited by J. Hu

77 Biologically inspired textiles
Edited by A. Abbott and M. Ellison

78 Friction in textiles
Edited by B. S. Gupta

79 Textile advances in the automotive industry
Edited by R. Shishoo

80 Structure and mechanics of textile fibre assemblies
Edited by P. Schwartz

81 Engineering textiles
Y. E. El-Mogahzy

Introduction

Textiles have one of the longest histories of all human activities. Beginning with prehistoric humans who fashioned clothing with available biological materials, such as animal skins and furs, the industry evolved to the eventual use of plant stalk derived fibers and seed hairs (flax and cotton). Depending on the geographic location, animal skins such as those from wolves, bears, deer, moose, and leopard were used as prized garment materials. In Chapter 8, the use of biomimetic principles applied to the structure and properties of animal furs in the fashioning of synthetic alternative textiles is discussed.

The use of animal skins did not require development of technology for conversion of fiber materials to fabrics; eventually, methods for construction of yarns from fiber were developed, and this led on to methods of basket weaving and conversion of yarns into fabrics. Concomitant with the rise in animal husbandry, significant amounts of animal hairs could be harvested from sheep (wool), goat, camel, and even rabbit and used as the feedstock for yarn and fabric. Similarly, with the increase in cultivation of plants, agrarian societies could produce substantial amounts of plant fibers to be used in textile manufacture. Thus, the fibers in the stalks of plants, leading to ramie, jute and hemp fibers, and seed hair of the cotton plant (perhaps viewed as advanced materials during that era) were relatively quickly incorporated into textiles.

An event that heralded one of the major advances of the new age of textile fibers, and which foreshadowed the work presented here in the opening chapters of this book, occurred in the mid-nineteenth century with the discovery of the solubility of nitrocellulose in simple solvents (e.g. acetone) allowing the production of synthetically derived fibers. Unfortunately, nitrocellulose, also known as guncotton, is highly flammable, so the commercial viability of these fibers was limited. Nonetheless, this concept illustrated that humankind was no longer strictly limited to the use of naturally produced material but was capable of producing novel fibers not present in nature. By the late nineteenth century, the method for production of cuprammonium rayon was discovered. This fiber, made from cellulose, was touted as being the manmade replacement for silk. During this same era, many other methods for converting cellulose from many natural sources were developed.

Another seminal event on the road to synthetic fiber production was the discovery by Carothers of the method for making nylon in the early twentieth century. This fiber is a truly synthetic material, in that the starting materials are not found in nature, albeit derived from petroleum. This discovery, and many more, was the genesis of the synthetic materials industry. As it relates to the current generation of synthetic materials, the power of human ingenuity evidenced by these developments in fibers is enormous.

Throughout this period many synthetic processes were investigated to replace natural silk production. Produced by silk moth caterpillars, and harvested through a labor-intensive process, silk was quite expensive; hence, it was deemed desirable to find a substitute process yielding a high value-added product of similar or higher quality. From the earliest guncotton fibers to present-day versions of regenerated cellulose, this quest has been followed, with varying degrees of success. Along the way, the lessons learned provided much insight into improving our capabilities for designing and producing novel materials with desired properties.

Design principles for clothing and for other applications of fiber-based textile materials are represented in several chapters in this work. In Chapter 6, Kapsali presents the use of biomimetic principles in the design of clothing. It should be noted, however, that, in today's world, textiles are not just about human clothing. Development of technical textile materials founded on biological principles is the subject of the two chapters (7 and 10) by Stegmaier in which self-cleaning materials, energy harvesting, and novel filtration materials, and others, are discussed.

There are many other applications of textiles such as composite reinforcement materials. Inspiration taken from biology for enhancement of these materials is discussed in Chapter 5 by Santulli. Although this is a relatively modern area, the use of clay-based composites in construction notwithstanding, this chapter brings the use of natural, sustainable (plant) fibers to the fore.

One of the main themes in this book is the use of molecular biology techniques for the production of fibrous materials. Chapters on these techniques form the genesis of the book. The scope and power of genetic engineering is the subject of much research, and the use of these techniques for fiber production will lead to as powerful a shift in our thinking as the early synthetic chemistry breakthroughs. In essence, the world of recombinant DNA technology has enabled us to refocus our attention on natural materials but through a distinctly new lens. In this book, we explore, via several contributions, the arena of biomaterials and the use of genetic engineering technologies as production systems to allow us not only to recreate extant materials but to create novel materials with new unexplored properties.

To realize the dream of producing designer materials from the products of simple living recombinant systems we need to understand the structure/function properties of the constituent natural ingredients of biomaterials (e.g. proteins and carbohydrates) and how natural systems process and produce biomaterials from these substrates. Thus, we present chapters on the examination of natural production systems for fibrous proteins and research into the processing steps involved in assembly of natural protein fibers. Further chapters deal with structural protein gene mimicry and production of recombinant proteins in simple microbial systems, and plant and animal systems.

This volume tells the complete story so far of biomimetics in textile manufacturing, leaving us with very exciting prospects for the manufacture of the next generation of novel textile materials for diverse applications using environmentally friendly technologies.

Part I

Biomimetic principles, production and properties

1

Recombinant DNA methods applied to the production of protein-based fibers as biomaterials

F. TEULÉ and R. LEWIS
University of Wyoming, USA

W. MARCOTTE JR and A. ABBOTT
Clemson University, USA

Abstract: The protein compositions of selected high-performance biological materials such as silks, collagen and mussel byssus threads are reviewed. The possible roles of key amino structural motifs present in these fibrous proteins are outlined. Experimental investigation of structure/function relationships in fibrous proteins is discussed. More specifically, the structural characterization of artificial silk proteins, produced through the expression of native silk cDNAs, synthetic silk gene analogs or designer silk genes in different expression systems, is reviewed in detail. Finally, the most popular and most successful unicellular or multicellular expression systems available for the production of such fibrous proteins are described.

Key words: expression of native or synthetic silk gene analogs, structural characterization, recombinant silk proteins, biological materials, expression systems.

1.1 Introduction

Biological organisms produce a wealth of natural fibrous polymers that have evolved to achieve highly specialized roles and to allow organism adaptation and survival. Some of the more interesting fibrous materials such as wool, silks, skin, cartilage and tendons are composed of fibrous proteins.

The literature is replete with details on the compositions, structures, mechanical and other physical and chemical properties of many of these protein-based biomaterials. This richness of information, along with scientific progress in the fields of genetic engineering and materials science, make it feasible to realize the production of biomaterials using recombinant DNA technologies.

It is the intent of this review to touch upon much of the rationale and methods for creation of genetically engineered systems for the production of fibrous protein polymers. We will focus on the utilization of spider silk gene mimics as a demonstration of a successful design, implementation and production scheme.

1.2 Biomimetics and protein-based biomaterials

Protein-based materials or protein fibers were thought for a long time to be a unique characteristic of the animal kingdom. However, such fibrous proteins have since been identified in plants. In wheat, a large, highly elastic storage protein called gluten, important in the expansion or rising of wheat-based doughs, was recently characterized as one of the first plant fibrous proteins (Tatham and Shewry, 2000).

The basic components of these natural fibrous materials are structural fibrous proteins. Such proteins are found in hair, tendons, cartilages, skins, arteries, muscles of mammals or in cuticles and silks of many arthropods. The individual proteins that make up these fibers or fibrous materials are of a specific sequence and often share structural amino acid motifs from one type of protein to another. In addition, many of these fibrous proteins have the ability to self-assemble, in an ordered manner, into a supramolecular network. The resulting supramolecular structure is often insoluble and is maintained, depending on its origin, by a combination of intra- and/or inter-molecular cross-linking, hydrophobic interactions, hydrogen bonding and coulombic interactions. Cross-linking is often necessary to hold the 'structure' together and can be different in nature depending on the amino acid residues involved. The molecular architecture (sequence and composition) of the individual proteins forming the network and the level and types of cross-links involved determine the mechanical and physical properties of the final fibrous materials, and the sequence in which these fibrous materials are assembled may be equally critical.

All these bio-based polymers display exceptional mechanical properties. More importantly, they are biodegradable and, thus, are attractive as a potential source of exploitable, environmentally-friendly materials. However, engineering new bio-based fibers having desired or customized mechanical properties is extremely challenging. To achieve this goal, the molecular architecture of the molecules composing these fibers, as well as their assembly processes, need to be fully understood in order to control assembly in the manufacturing process. Protein-based fibers are very suitable for bioengineering and remarkable efforts are being made in this field to understand the structure/function relationships of protein polymers. Because of the availability of technologies allowing the manipulation of genes encoding proteins, the prospect of designing and manufacturing new, possibly customized, protein-based polymers seems within reach.

Although there are many diverse types of fiber proteins, we will limit our discussion primarily to silk and collagen proteins for the sake of brevity. However many of the rules that govern the structure/function relationships of these specific fibrous materials or fibers are relevant to others as well (e.g. elastins-, lamprin-based materials).

1.3 Characteristics of some natural protein-based materials

1.3.1 Homopolymers

Silks

Many arthropods such as insects (Lepidoptera), and arachnids (Araneidae) produce silks. These highly insoluble proteinacious fibers share several structural motifs and characteristics. The primary structures of the different silk proteins are usually highly repetitive and comprise combinations of crystalline structures (tightly packed β-pleated sheets with more loosely associated β-sheets) and amorphous regions (α-helices, β-spirals or 'spacer' regions) (Xu and Lewis, 1990; Beckwitt and Arcidiacono, 1994; Guerette *et al.*, 1996; Simmons *et al.*, 1996; Hayashi and Lewis, 1998). Although most silk-producing organisms may only produce one type of silk, some arachnids such as Araneidae spiders have a more complex spinning apparatus and are able to produce several types of silks (Peters, 1955; Lucas, 1964). In these spiders, the different silks originate from different silk glands and differ in both amino acid composition and mechanical properties. The wide range of mechanical properties exhibited by these different silks allow for their adapted use in various applications from web construction, prey swathing, safety line or dragline, to cocoon construction (Lewis, 1992).

The cocoon silk of moth larvae such as *Bombyx mori* (Bombycidae, Lepidoptera) is the main source of silk exploited by textiles industries. *B. mori*, like all lepidopteran caterpillars, possesses a pair of modified salivary glands (labial glands) that are responsible for silk production. This silk-secreting system is linked to an outlet by which the silk is pulled out as the worm moves its head from left to right following a figure-eight-shaped trajectory (Sehnal and Akai, 1990).

The fibrous core of *B. mori* cocoon silk is composed of a 350 kDa heavy-chain fibroin (H-fibroin, *Bm-Fhc*) and a 25 kDa light-chain fibroin (L-fibroin, *Bm-Lc*) covalently linked by disulfide bonds. These fibroins are also associated with a small glycoprotein of approximately 27 kDa (P25) but there is no evidence of covalent links (Tanaka *et al.*, 1999a; Tanaka *et al.*, 1999b). P25 may play a role in the transport of the two fibroins from the cells to the gland lumen. The heavy-chain fibroin is the fiber protein responsible for silk formation.

B. mori silk fibroins are characterized by a high glycine content. The fibroin primary sequence is highly repetitive and displays a noticeable $(GA)_n$ motif containing glycine and alanine residues interspaced with serine-containing polyhexapeptide repeats GAGAGS and GAGAGX with X = Y or V (Zhou *et al.*, 2000). The short GAGAGS repeat can form crystalline regions in both morphological states of silk studied: silk I and silk II. Silk I or 'water soluble silk' is obtained from the desiccated gland content without any mechanical disturbance (storage state of the fibroin) (Lotz and Cesari, 1979). The more stable silk II (β-silk fibroin) is observed in the spun cocoon fiber. Silk I is converted to silk II by mechanical stress or shearing that occurs when the fiber is being pulled out. The crystalline structure of silk II is viewed as extended polypeptide chains arranged in hydrogen-bonded structures. In this structure, the GAGAGS motif adopts an anti-parallel β-sheet so that all the glycine residues are projected on one side of the sheet and all alanine or serine residues are projected on the other side. However, in silk I, these β-sheets are imperfect (Lotz and Keith, 1971; Fossey *et al.,* 1991).

In *B. mori* silk, crystalline regions account for approximately 40 to 50% of the silk structure. Typically, the fibroins consist of twelve large crystalline domains. These crystalline domains are periodically interspaced with short distinct 'spacers' (sequences with bulky amino acid residues like Y, W, E and R) that are more amorphous regions with random orientation. These spacers are important because they may give some flexibility to the fiber (Denny, 1980). The sticky protective coating of *B. mori* silk is composed of a protein called sericin. This 'glue' has a critical role since it provides proper adhesion of the cocoon fibers with one another (Grzelak, 1995).

B. mori silk has a very specific function since it is used by the caterpillar to build the cocoon in which it pupates. As a result, to ensure proper development of the pupae, the silk has to be very strong and resilient; elasticity in this case can be considered a luxury. The mechanical properties of *B. mori* silk are a tensile strength of 0.5–0.74 GPa, an elastic modulus (Young's modulus) of 5–17 GPa and an extensibility of 19–24% (Denny, 1980; Cunniff *et al.*, 1994; Perez-Rigueiro *et al.*, 2000).

The most interesting spider silks are those produced by the orb weavers *Nephila* and *Araneus* (Araneidae). These spiders possess seven types of glands and are able to produce six types of silks (and a glue) with very different mechanical properties (Denny, 1976; Gosline *et al.*, 1984; Gosline *et al.*, 1999; Gosline *et al.*, 2002) and amino acid compositions (Andersen, 1970; Work and Young, 1987; Lombardi and Kaplan, 1990). Each has a determined function and use (i.e. web building silks, cocoon silk, dragline and swathing silk). The most fascinating and thus best-characterized silks are the dragline or major ampullate silk and the flagelliform or viscid silk. Dragline silks are used for construction of the web frame or as a safety line when the orb weaver spider drops from high elevations and flagelliform silks make up the capture spiral of the web. These silks differ in their repeat structures resulting in different tensile strength and elasticity.

Studies on several major ampullate dragline silks revealed that this fiber is composed of two proteins called spidroins (spider fibroins, MaSp 1 and MaSp 2) that contain glycine-rich and polyalanine repeats (Xu and Lewis 1990; Hinman and Lewis, 1992). X-ray diffraction and NMR spectroscopy studies demonstrated that the polyalanine regions are organized in anti-parallel β-sheets alternating with the glycine-rich amorphous regions (Hirijida *et al.*, 1996; Simmons *et al.,* 1996). The polyalanine domains can form non-covalent cross-links or hydrogen bonds between the individual proteins and alternate either with a glycine-rich tripeptide domain [$(GGX)_n$, MaSp 1] or with a glycine-rich, proline-containing pentapeptide [$(GPGXX)_n$ = GPGGY or GPGQQ, MaSp 2]. The GPGXX motif is predicted to be responsible for the fiber's elasticity (Guerette *et al.,* 1996; Hayashi and Lewis, 1998) and the GGX motif, which can form a small 3_{10} helix or more likely a glycine II (gly II) helix, is predicted to link the highly crystalline and more amorphous regions (Hayashi *et al.,* 1999). The structure of this GPGXX pentapeptide motif resembles the elastin pentapeptide VPGVG and thus may form β-turns. Tandem repetitions of these structures would yield an elastic β-spiral and act as a spring (Hayashi *et al.,* 1999).

Flagelliform silk is rich in glycine and proline (Andersen, 1970) and characterized by the presence of long elastic domains containing a GPGGX elastic motif similar to that found in dragline silk proteins, interspaced with GGX tripeptides. The GGX tripeptides always precede non-repetitive 'spacer' regions that are thought to be involved in cross-linking of the silk fiber (Hayashi and Lewis, 1998; Hayashi and Lewis, 2000). In contrast to dragline silk proteins, flagelliform silks do not contain polyalanine repeat motifs.

Both dragline and flagelliform silks behave like viscoelastic materials. Spider dragline silks (major ampullate silks) are very tough fibers (160 MJ m^{-3}) due to a combination of high tensile strength (1.1–4 GPa), high stiffness or elastic modulus (10 GPa), and high extensibility (35%) (Denny, 1976; Gosline *et al*., 1984; Gosline *et al*., 1999; Gosline *et al*., 2002). Like dragline silk, flagelliform silks are also very tough fibers (150 MJ.m^{-3}), but they are far more elastic than dragline silks (200–270%) and, thus, have a lower stiffness (3 MPa) (Denny 1976; Gosline *et al*., 1984; Stauffer *et al*., 1994). Their extreme elasticity seems to be the result of the marked presence of these long regions containing repeats of several elastic motifs (Guerette *et al*., 1996; Hayashi and Lewis, 1998; Hayashi *et al*., 1999). The mechanical properties of this viscid silk are similar to those of lightly cross-linked rubbers (Gosline *et al*., 1986).

A phenomenon called supercontraction has been observed for some spider silks. Although at room temperature *Araneus diadematus* dragline silk is very stiff, immersion of that silk in water causes a contraction to about 55% of its dry length, a diminution of the elastic modulus and an increase in extensibility (Gosline *et al.,* 1984).

Collagens

Collagens represent one of the most abundant proteins found throughout the entire animal kingdom. Most of the time, these highly insoluble fibers are used to support the framework of many body parts as stress-bearing components. For instance, skins or leathers, cuticles, and connective tissues such as cartilages, teeth, tendons, ligaments, blood vessels and bones contain collagen fibers.

The extracellular protein matrix consists of self-assembled tropocollagen molecules synthesized in specialized cells (fibroblasts). Usually, the individual tropocollagen α-chains contain a GXY tripeptide repeat in which X and Y are bulky and inflexible proline and hydroxyproline residues, respectively. Each of the individual α-chains adopts a left-handed helix and three helical chains twist together to form the tropocollagen right-handed triple helix. Intermolecular hydrogen-bonding stabilizing the triple helix may occur between glycine and the carbonyl of the X residue present at the same level.

1.3.2 Copolymers: marine mussel byssus threads

Marine mussels (*Mytilus edulis*) have the unique ability to generate protein-based tendons or byssus threads to bond to solid surfaces in wet, saline, and turbulent environments (Pujol, 1970; Waite, 1992). Bonding is rapid and it only takes a mussel five minutes to generate a 2–4 cm thread by injection molding (Waite, 1992). This type of bonding is also permanent and extremely versatile. Such byssus threads or 'beards' are extra-corporeal bundles of tiny tendons which are attached distally to a foreign substrate such as a rock and proximally by insertion of the stem root into the byssus thread retractor muscles.

Unlike regular tendons, byssus threads have a non-periodic microstructure. Two collagen-like polypeptides called Col-P and Col-D, both homotrimers (50 and 60 kDa), exist in complementary gradients along the length of the byssus thread. Col-D dominates at the distal end of the thread or close to the adhesive plaque and Col-P dominates at the proximal end or close to the shell. Both precursors of Col-D and Col-P, called preCol-D and preCol-P, were identified in mussel foot extracts (Coynes *et al.*, 1997; Qin *et al.*, 1997). A third protein, preCol-NG, may mediate the graded progression of preCol-P and preCol-D between proximal and distal ends (Qin and Waite, 1998).

The preCol-P protein was the first described natural block copolymer containing a hard collagenous domain flanked on either side by soft elastin-like domains. It contains a central region of 435 residues corresponding to the collagenous domain or 146 (GXY) repeats (X = proline or *trans*-hydroxyproline; Y = *trans*-4-hydroxyproline). The (GXY) repeat is interrupted after the eleventh repeat GST by a single missing glycine (G) causing a kink in the molecule and the bending of the triple helix (Coynes *et al.*, 1997). An acidic patch of 15 amino acids rich in glutamic acid (E) and aspartic acid (D) is located after the collagen sequence. Elastic domains flank the collagen domain on both sides and could provide solubility to the protein. These domains, dominated by glycine, proline (P) and bulky hydrophobic residues are 'proline-containing pentapeptide repeats' (XXXPX). This pentapeptide resembles the elastin VGVPG pentapeptide in which glycine, proline and the hydrophobic amino acids of elastin are critical for the entropically driven elastic recoil (Coynes *et al.*, 1997).

The primary structure of the predicted preCol-D protein consists of four domains symmetrically arranged around a 45.5 kDa collagenous domain (= 175 GXY repeats) that is identified as a type-III collagen because of the high level of glycines. These repeats more closely resemble a GGX repeat rather than a GXY repeat. This GGX tripeptide destabilizes the triple helix and thus may give more flexibility to the chain in the affected regions (Qin *et al.*, 1997). There are also differences from regular collagens such as glycine substitutions and residue deletions or the presence of unusual amino acids such as tyrosine (Y) and cysteine (C). Apparently, these breaks in the collagen continuity form stable bends or kinks in the triple helical structure since they are resistant to high pepsinization. As for preCol-P, there is a stretch of 20 hydrophobic and acidic amino acids at the end of the collagenous domain and there is also a proline-rich hinge region of 44 residues. Flanking the central collagenous domain of preCol-D are silk fibroin-like domains (dragline silk spidroin 1-like) containing several polyalanine motifs of 11–14 amino acids (2 and 5 repeats in the N- and C-termini, respectively) and GGX tripeptide repeats. Under shear, these polyalanine segments may form β-sheets with expanded sheet spacing because of various amino acid substitutions and the presence of bulky amino acids (R, Q).

Like preCol-D and -P, preCol-NG contains a central collagenous domain, flanking domains, an acidic patch and histidine-rich termini. Its flanking domains resemble the glycine-rich proteins of plant cell walls with tandem $(XG)_n$ repeats (X = A, L, or N but not P). PreCol-NG also has sequences similar to the (GA) and polyalanine runs found in arthropod silks. This protein seems to act as a mediator between preCol-D/-P molecules during assembly (Qin and Waite, 1998). The byssus thread fibers are highly cross-linked. N- and C-termini of preCol-P, -D, and -NG contain histidine-rich regions which may be involved in cross-linking through metal (possibly zinc) complexing (Coynes *et al.*, 1997; Qin *et al.*, 1997). Dityrosine bonding or aryl grouping involving tyrosine (Y) residues in preCol-P and preCol-D also stabilize end-to-end adhesion (Waite *et al.*, 2002).

Each byssus thread is a stiff strong tether at one end and a shock absorber at the other end owing to the graded distribution of tensile molecular elements. Collagen domains may confer self-assembly to all parts of the structure while histidine-rich regions provide cross-linking sites resulting in metal-induced stickiness holding the structure together. The graded distribution of both proteins resulting in the graded axial distribution of elastic elastin-like and stiffer silk-like flanking domains seems critical to moderate the stress concentration of this highly composite structure (Waite *et al.*, 2002). These threads, which are five times tougher than Achilles tendons, can shrink and have a melting temperature exceeding 90 °C. The presence of the elastin blocks in the Col-P protein may improve the extensibility of the collagen and its tensile stress. As a result, the material exhibits a breaking strength lower than that of tendons (35–75 MPa) but is much tougher than tendons (35–45 MJ. m^{-3}) because it is more extensible (109–200%; Bell and Gosline, 1996).

1.4 Experimental characterization of model fibrous proteins

The characterization of natural fibrous proteins has shown that many are highly repetitive and contain similar short amino acid structural motifs. The sequence and order of these structural motifs (primary sequence and repeat pattern) impact folding of the protein chain as well as further interaction of the protein with itself and/or others. Therefore, in a macromolecular polymeric material formed by the assembly of similar structural protein chains, those amino acid repeats may be responsible for the physical and mechanical properties of the polymer. It is currently possible to produce proteins of particular specific amino acid sequence and length and there is a great interest to use these technologies to produce sufficient fibrous protein materials for research because of the huge potential for the development of 'bio-inspired or biomimetic materials'.

1.4.1 Production of fibrous protein analogs

Overall, biophysical and chemical data on spider silk (x-ray diffraction, FTIR, NMR and CD) provide a limited understanding of the mechanical properties of spider silks (Hirijida *et al.*, 1996; Simmons *et al.*, 1996; Parkhe *et al.*, 1997; Dicko *et al.*, 2004). The ideal situation would allow the use of full-length, native fibrous proteins for direct assessment of structure/function relationships and assembly mechanisms. However, because of the highly repetitive nature and frequent large size of these polypeptides, this has not been possible. As a result, recombinant DNA technologies have been exploited as a way to produce significant amounts of native or tailored fibrous proteins to further investigate these relationships and gain a better understanding of how these molecules perform in the natural protein polymers, including mechanisms of protein self-assembly. In addition, analogs of structural proteins may yield materials with novel properties. Most protein polymers engineered so far have been modeled on natural structural proteins such as elastin, silks or collagen. To be concise, we will focus on the studies relating the genetic engineering of silk-like analogs.

Expression of native silk cDNAs

Several partial dragline silk cDNA sequences have been cloned and expressed to produce native silk analogs. A partial MaSp 1 cDNA from *Nephila clavipes* (1.5 kbp) was expressed in *Escherichia coli* (Arcidiacono *et al.*, 1998) and the 43 kDa His-tagged (polyhistidine-tagged) recombinant protein was produced *in vivo* at low levels (~4 mg of purified protein L^{-2}). However, the presence of truncated proteins was evident on Western-blot analysis suggesting some level of gene recombination in *E. coli*. Other cDNA sequences encoding dragline silk proteins from two spider species (*N. clavipes* MaSp 1 and MaSp 2; *Araneus diadematus* ADF-3) were cloned and expressed in two mammalian cell lines, bovine mammary epithelial alveolar cells/MAC-T or baby hamster kidney cells/BHK (Lazaris *et al.*, 2002). Multimers of the ADF-3 cDNA sequences were also engineered. The secreted silk proteins, ranging from 60 to 140 kDa, were soluble in phosphate saline buffer, possibly owing to the presence of the native hydrophilic C-terminus, and easily purified. The larger the protein, however, the lower the protein yields. Protein production levels reached 25–50 mg/L^{-1} in BHK cells for the ADF-3 monomer gene and synthetic fibers were obtained by wet spinning (Lazaris *et al.*, 2002). A more recent study reports the expression of small flagelliform and MaSp 1 cDNAs (*N. clavata*) into BmN insect cells (*B. mori*-derived cells; Miao *et al.*, 2006; Zhang *et al.*, 2007). This MaSp 1 cDNA was also expressed in transgenic silkworm larvae (Zhang *et al.*, 2007). The enhanced green fluorescent protein (EGFP)-MaSp 1 protein (70 kDa) was highly insoluble and had a tendency to aggregate. The recombinant silk protein yields were only 5% of the total proteins in cell culture and 6 mg per transgenic larvae (Zhang *et al.*, 2007).

Expression of synthetic silk analogs

Many laboratories also copied consensus repeat sequences of native dragline silk proteins (*N. clavipes* MaSp 1 and MaSp 2) to engineer synthetic silk-like gene replicates for the production of spider silk-like proteins in different expression systems. Synthetic MaSp 2 genes (*N. clavipes*) containing 8–32 repeats of a monomer unit (105 bp) encoding a PGGYGPGQQGPGGYGPCQQGPSGPGS $(A)_8G$ consensus sequence were cloned in *E. coli* (Lewis *et al.*, 1996). The recombinant MaSp 2 protein (16 repeats) was produced at level of 2–10 mg/g^{-1} of wet cells. The pure lyophilized silk protein was dissolved in formic acid (2–4 mg/ml) and extrusion of the protein solution into methanol yielded a solid fiber thread (Lewis *et al.*, 1996). Native MaSp 1 or MaSp 2 sequences (*N. clavipes*) were also used to construct larger inserts using bigger synthetic DNA monomer sequences (303 bp MaSp 1 and 357 bp MaSp 2 monomers) (Fahnestock and Irwin, 1997). Three MaSp 1 synthetic variants (each with native deletion patterns) were constructed. The synthetic MaSp 2 sequences constructed, about 4.8–5.6 kbp (up to 16 monomer tandem copies), were cloned in *E. coli* (Fahnestock and Irwin, 1997). Additional synthetic MaSp 1 genes containing 8–16 repeats of the 303 bp monomer sequences were constructed and cloned in yeast (*Pichia pastoris*) (Fahnestock and Bedzyk, 1997). Using the recombinant genes described above, silk proteins of different sizes were produced. Although longer genes encoding proteins containing up to 3000 residues could be expressed in *P. pastoris,* the protein yields dropped drastically. In contrast to *E. coli,* yeast stable integrants did not show any evidence of truncated synthesis or gene recombination (Fahnestock and Bedzyk, 1997).

Synthetic silk-like genes were also expressed in transgenic tobacco and potato (Scheller *et al.*, 2001). The synthetic spidroin 1-like sequence, engineered by ligation of multiple sequence cassettes flanked by appropriate restriction sites, thus taking into account that the MaSp 1 repeats have slight differences, replicated almost exactly the native MaSp 1 sequence. The 1.8 kbp synthetic MaSp 1 construct, placed under the control of the CaMV 35S viral promoter lacked the native 3' non-repetitive terminus but contained an N-terminal LeB (legumin) signal sequence to direct the protein into the secretory pathway, a C-terminal ER retention signal (KDEL) and a c-myc tag for immunological detection. MaSp 1 proteins (12.9–99.8 kDa) were produced at a level of 2% of the total soluble protein in tobacco leaves or potato tubers (Scheller *et al.*, 2001).

1.4.2 Production of designer fibrous protein analogs

Several laboratories engineered synthetic spider silk-like gene constructs to specifically investigate the mechanisms of interaction of the individual silk molecules as well as the structural conformations taken by identified key silk amino acid structural motifs. In some cases, the gene engineering, which usually targets protein self-assembly, includes self-assembly trigger sequences in order to better manage protein aggregation.

Homoblock protein polymers made of silkworm silk-like crystalline and flexible blocks were produced in *E. coli* (Capello *et al.*, 1990). DNA sequences encoding the short repeat motifs corresponding to the crystalline region [poly(GAGAGS)] of the *B. mori* silk fibroin alone or containing a non-crystalline region from silk (GAAGY) were first cloned in *E. coli* as monomer units (Capello *et al.*, 1990). Silk-like protein polymers (SLP) were produced and purified from bacteria using a highly concentrated lithium bromide solution (LiBr). Another group designed an additional copolymer, 'SLPF', containing the silk-like repeat GAGAGS juxtaposed to a fibronectin sequence GAAVTGRGDSPASAAGY which was produced in *E. coli* (Anderson *et al.*, 1994). The SLPF protein $[(GAGAGS)_9$ $GAAVTGRGDSPASAAGY]_{12}$, had the physical stability of silk and the cell attachment bioactivity of human fibronectin given by the (RGD) triplet. The structures of lyophilized SLP and SLPF proteins were analyzed by wide angle x-ray diffraction scattering (WAXS) and transmission electron microscopy (TEM). WAXS data supported the theoretical x-ray diffraction pattern indicating that dry SLP and SLPF proteins accommodate the same anti-parallel β-sheet crystal structure characteristic of native silks due to their silk component. The overall crystallinity of the SLP polymers is affected when adding non-crystallizing blocks (i.e. GAAGY) to their sequence. The SLPF polymer is semi-crystalline. TEM studies showed that the fundamental unit of the morphology of SLPF is a whisker crystal that can form sheaves (Anderson *et al.*, 1994). The SLP and SLPF polymers were solubilized at concentrations up to 400 mg/ml in LiBr. These synthetic polymers solutions were also spun into fibers.

Another synthetic multiblock ('polymer 1') containing sequences inspired from dragline silk protein sequences was produced in *E. coli* (Qu *et al.*, 2000). The 'polymer 1' produced $((AEAEAKAK)_2AG(GPGQQ)_6GS]_9(AEAEAKAK)_2AG$ (GPGQQ) could spontaneously form a self-supporting macroscopic film as a result of conformational transition (α-helices to β-strands) of segments within the polypeptide. FTIR, CD spectroscopy and ^{13}C NMR showed that the alanine-containing segments formed α-helices in solution and β-sheets in films (Qu *et al.*, 2000).

Winkler *et al.* (1999, 2000) expressed small genetically engineered spidroin variants (25 kDa) containing methionines (M) or RGYS*L trigger sequences recognized by a cyclic AMP-dependent kinase (cAPK) flanking a penta-alanine motif in *E. coli.* Once oxidized, the methionine residues acted as sterical triggers to prevent β-sheet assembly, thus allowing control over the protein assembly process. The RGYS*L trigger present in the same type of protein served to study the effect of phosphorylation/dephosphorylation on the recombinant protein β-sheet self-assembly capacity (Winkler *et al.*, 2000). Phosphorylation of these analogs on their trigger sequences increased their solubility by disrupting the β-sheets. These approaches demonstrate clever ways to control protein self-assembly chemically or biologically, thus preventing the premature aggregation of genetically engineered silk-like proteins.

Other studies focused on investigation of the structure adopted by the spider silk-like (GGX) motifs. Synthetic MaSp 1 homopolymers or MaSp 1/MaSp 2 block copolymers of variable sizes were expressed in *E. coli* and the His-tagged proteins ranging from 14.7 to 41.3 kDa were purified by IMAC (immobilized metal affinity chromatography) followed by reverse-phase HPLC. CD data on the recombinant silk proteins indicated that two-thirds of the modular proteins adopted β-sheet-structures and the remaining one-third was present as β-turn structures (Prince *et al.*, 1995). These results suggested that the GGX motifs from the MaSp 1 sequences adopted β-sheet structures (Prince *et al.*, 1995). In a later study, a sequence encoding a SCAP 1 polypetide (GLGGQGGGAGQGGYG) was engineered and cloned in *E. coli* (Fukushima, 1998). The SCAP 1 sequence was then condensed to form the SCAP multimers containing 4 to 13 tandem repeats of SCAP 1. All the SCAP sequences were expressed in *E. coli* and purified by IMAC. Typically, protein yields in fermenters decreased with increased molecular weight of the produced proteins (from 5.2 to 1.2 mg/L). Study of lyophilized SCAP 13 (13 repeats) suggested that the glycine-rich region of the protein adopted random coils whereas as in a film cast from formic acid, it adopted β-sheet structures. These results also indicated that the glycine-rich sequence present in MaSp 1 (GGX) adopted β-sheet structures (Fukushima, 1998).

These results, though in agreement with previous data collected by TEM on spider dragline silk fibers (Thiel *et al.*, 1994) are in total contradiction with NMR (Kümmerlen *et al.*, 1996), FTIR (Dong *et al.*, 1991) and REDOR-NMR studies (rotational echo double resonance) (Jelinski, 1998) that all suggested that these GGX segments could not form β-sheets and instead assumed compact helix-like structures. Others also suggested that these GGX segments most likely took on helix-like structures with three amino acids per turn (3_{10}) (Hayashi *et al.*, 1999).

A different MaSp 1/MaSp 2 hybrid sequence was engineered and expressed in *E. coli* (Ouroudjev *et al.*, 2002). This sequence comprised four MaSp 1-like sequences fused to one MaSp 2-like domain [pS(4+1)]. The individual MaSp 1 sequence (38 amino acids) encoded a 16 amino acid hydrophobic segment (polyalanine/(GA)), flanked on each side by 22 amino acid glycine-rich regions $(GGX)_{22}$. The MaSp 2 sequence encoded the $(GPGGX)_2$ polypeptide. Atomic force microscopy (AFM) studies on the synthetic pS(4+1) proteins revealed that the synthetic soluble recombinant proteins spontaneously formed nanofibers characterized by a segmented substructure (Ouroudjev *et al.*, 2002). The model proposed in order to explain the morphology of these nanofibers suggested that the protein molecule folded on itself via arrangement of the four MaSp 1 hydrophobic polyalanine/(GA) segments into anti-parallel β-sheets (crystals) stabilized by hydrogen bonding. The more hydrophilic $(GGX)_n$ segments of MaSp 1 assumed random coils or 3_{10} helices that alternated with the hydrophobic regions. The MaSp 2 sequences formed loops allowing the molecule to fold back onto itself. Thus, in this model, each folded molecule resembles a slab composed of alternating hydrophobic crystalline regions with more hydrophilic glycine-rich regions. Approximately 30 folded slab-like molecules form a 'stack' or 'nanofiber segment' that piles up through hydrophobic interactions of their alanine-containing segments (Ouroudjev *et al.*, 2002). When stretched or drawn, as in the natural silk spinning process, the $(GGX)_n$ segments composing the semi-amorphous matrix of the fiber may undergo a conformational change to more extended structures (3_1 helix or β-sheet) stabilized by newly formed intra- and inter-hydrogen bonds (Ouroudjev *et al.*, 2002).

Additional studies focused on the role and structure of the flagelliform silk-like $(GPGGX)_n$ motif (with X = A, V, Y, or S) that differs slightly from dragline silk MaSp 2 $(GPGX_1X_2)_n$ motif (with X_1X_2 = QQ/GY). The first study produced the recombinant 'polypetide 1' $[(GPGGSGPGGY)_2GPGGK]_{11}$ in *E. coli* (Zhou *et al.*, 2001). Structural data of the purified polypeptide 1 analyzed by CD, FTIR and NMR techniques indicated that the GPGGX motif formed small amounts of β-turns in solution and higher amounts in films (dehydrated polypeptide). In a more recent study, two synthetic silk proteins variants ($A1S8_{20}$ and $Y1S8_{20}$, each 60 kDa) were produced in *E. coli* to study the impact of a single amino acid substitution (**X**) in the (GPGG**X**)$_2$ motif on protein structure and self-assembly (Teulé *et al.*, 2007). The $A1S8_{20}$ and $Y1S8_{20}$ proteins contained 20 repetitions of sequences composed of the flagelliform-like motif ([GPGG$\mathbf{X}_1$ GPGG$\mathbf{X}_2$]$_2$, ($\mathbf{X}_1$/$\mathbf{X}_2$ = A/A or Y/S for the 'A1' or 'Y1' elastic versions, respectively) linked to an MaSp 2-like 'S8' motif {= $[GGPSGPGS(A)_8]$}. Both $A1S8_{20}$ and $Y1S8_{20}$ were heat-stable. CD data indicated that, in aqueous buffers, both proteins went through a heat-inducible β-sheet transition (attributed to the S8 hydrophobic crystalline-forming segment) that was more prominent and only irreversible for $Y1S8_{20}$. The more hydrophobic A1 motif (GPGG**A**)$_4$ present in the $A1S8_{20}$ protein seemed to favor a more random coil conformation with few 'loose' β-sheets in solution until the reversible structural transition of the S8 motif was induced by heat. The more hydrophilic Y1 motif (GPGG**Y** GPGG**S**)$_2$ favored a higher initial amount of stable β-sheet that increased during heating, albeit irreversibly. The A1 and Y1 motifs seemed to adopt different conformations in solution (random coil and possibly β-turns, respectively). Self-assembly of the pure proteins in aqueous environments was spontaneous for $Y1S8_{20}$ and shear-induced for $A1S8_{20}$ but, in both cases, the surface liquid films formed could be used to generate artificial silk fibers of different properties (Teulé *et al.*, 2007).

1.5 Expression systems available for recombinant fibrous protein production

With the advent of molecular biotechnologies, recombinant protein production is now possible in a variety of prokaryotic and eukaryotic systems. The choice of the host system depends on several factors such as the nature of the protein to be produced (requiring significant *vs.* little post-translational modifications), the purpose of its production (structural characterization *vs.* large-scale production), the mode of production (intracellular *vs.* secreted), and the method of purification. In all cases, a chimeric gene, containing sequences encoding for the desired protein, is introduced in the multicloning site of a plasmid expression vector that is introduced in the host system by transformation (chemical, electroporation, bacterial or biolistically mediated). The vector features required for proper gene expression depend on the host expression system but the simplest ones generally contain (1) a multicloning site, (2) promoter sequences that most of the time are chemically inducible or viral in origin (strong promoters), (3) appropriate 5'UTR/3'UTR flanking the sequence of interest, (4) a selectable trait (usually antibiotic resistance gene) allowing the selection of the transformants, and (5) an appropriate origin of replication for replication of the vector if the gene is not integrated in the host genome.

1.5.1 Prokaryotes

The simplest of all systems is bacteria, in particular *E. coli.* The advantages of bacterial systems include relative ease of handling, rapid rate of reproduction in simple liquid media, and tolerance to harsh treatment conditions. In addition, there are a large variety of inducible expression vectors, specialized bacterial strains that accommodate rare codons leading to enhanced expression levels, and multiple options for affinity purification of recombinant proteins (Makrides, 1996; Hannig and Makrides, 1998; Baneyx, 1999; Hunt, 2005). Moreover, recombinant bacteria grown in bioreactors allow mass production of the desired proteins.

For example, in this type of system, the coding region for a specific protein can be cloned into a plasmid vector under the control of a promoter juxtaposed to a lac operator (part of the lactose operon). In the absence of lactose, a repressor is bound to the operator preventing binding of RNA polymerase to the promoter region thus negatively regulating transcription. Addition of the lactose analogue IPTG (isopropyl-β-D-thiogalactopyranoside) to the culture media causes release of the repressor and thus transcription of the gene and protein production. Addition of sequences encoding a removable histidine-tag located at the amino or carboxy terminus of the chimeric protein facilitates purification of the proteins by IMAC (Hochuli *et al.*, 1987; Hochuli, 1988; Hochuli, 1990).

However, *E. coli* cells have been reported to have difficulties handling repetitive DNA sequences (recombination and deletion). In some cases, truncated products were observed even when expressing relatively short silk-like polypeptides (Arcidiacono *et al.*, 1998). The *E. coli* system was also depicted as limited because of the observed premature termination errors of protein synthesis when attempting to produce proteins containing more than 1000 amino acids (Fahnestock and Irwin, 1997). In other cases, genetic instability of larger silk-like inserts resulted in internal deletion or duplication resulting in the generation of a ladder of protein products (Fahnestock and Irwin, 1997). Further, some studies showed that even though the codon used to engineer synthetic silk genes was optimized to maximize expression levels in *E. coli,* recombinant protein yields were inversely proportional to the length of the multimer sequence (Prince *et al.,* 1995).

As prokaryotes, bacteria are also limited in the type of post-translational modifications that can be performed. While disulfide cross-links can be made, bacteria are unable to perform typical eukaryotic post-translational modifications (secretion signal processing, O- and N-glycosylations, or lipid additions). As a result, in certain cases the protein formed is often improperly folded.

1.5.2 Eukaryotes

Unicellular systems: yeast

Yeast is a simple single-celled eukaryotic organism, which, like bacteria, has a rapid rate of reproduction, can be propagated in simple liquid media in bioreactors, and is suitable for large-scale protein production. The simplicity of the techniques used to genetically manipulate yeast makes this system extremely tractable for eukaryotic expression of designer proteins. Expression vectors are available for both *Saccharomyces cerevisiae* (baker's yeast) and *P. pastoris* (methylotrophic yeast). Moreover, the characterization of secretion signals in *S. cerevisiae* such as the 87 amino acid MAT α-prepro signal peptide (Lin and Cregg, 2001) and its engineering in yeast expression vectors allow the secretion of the produced recombinant protein outside the cell. All secretion signals are cleaved upon targeting of the protein to the secretion pathway and thus do not interfere with the native secondary structure of the protein. Protein secretion is also an option with *P. pastoris*, although secretion of native proteins occurs at lower levels compared with *S. cerevisiae* (Cereghino *et al.*, 2002).

Despite the differences in secretion efficiency, the *P. pastoris* system has the advantage of permitting very high culture densities in bioreactors. Designed vectors for the *P. pastoris* system (e.g. Invitrogen) are shuttle/expression vectors for easy gene engineering in *E. coli* with subsequent transfer to yeast for protein production (intracellular or extracellular). Upon transformation, the linearized recombinant plasmid becomes inserted in the yeast genome by homologous recombination, usually by a single cross-over event, at the alcohol oxidase 1 (AOX1) gene locus.

The sequence of interest is thereby placed under the control of the AOX1 gene promoter and its expression in yeast is induced by the addition of methanol to the culture media (Cregg and Madden, 1988). The native AOX enzyme, required for methanol metabolism, accumulates intracellularly in the peroxisome, where it catalyzes the formation of formaldehyde using methanol as a substrate. Some of the formaldehyde leaves the peroxisome and is further oxidized to generate energy for the growth of this organism while the remaining is integrated in a cyclic pathway to generate cell constituents (Cereghino and Cregg, 2000). The levels of accumulation of AOX can reach 30% of the soluble protein fraction in cells grown in fermenters supplied with methanol. Thus, the *P. pastoris* system produces more significant amounts of protein intracellularly and extracellularly than other yeasts.

Many eukaryotic post-translational modifications are also possible in yeast. For instance, yeast produces at least two kinds of amine oxidases differing in heat stability and substrate specificity, methylamine oxidase and benzylamine oxidase. A *P. pastoris* benzylamine oxidase was characterized and, although this 120 kDa enzyme displayed a wide range of substrates, there was a clear preference for peptidyl lysine. This enzyme was named PPLO for *P. pastoris* lysyl oxidase (Kuchar and Dooley, 2001). Thus, *P. pastoris* is able to cross-link proteins through lysyl residues. Other enzymes such as protein disulfide isomerases (PDI) are also produced in *P. pastoris* (Warsame *et al.*, 2001) and, thus, protein cross-linking through disulfide bonds is possible. *P. pastoris* is also able to perform post-translational modifications such as glycosylation. Indeed, evidence for the presence of O- and N-linked carbohydrates to several recombinant proteins was observed (Cereghino and Cregg, 2000). However, the glycosylation mechanisms are not well known and some foreign proteins that are not extensively glycosylated in their native host sometimes end up hyperglycosylated in *P. pastoris*.

This system also seems to be more tolerant in handling repetitive DNA sequences (e.g. silk-like sequence) compared with *E. coli*. In contrast to *E. coli*, stable yeast integrants do not show any evidence of truncated synthesis or silk-like gene recombination (Fahnestock and Bedzyk, 1997). However, although longer silk-like synthetic genes encoding proteins containing up to 3000 residues could be expressed in *P. pastoris*, the protein yields dropped dramatically (Fahnestock and Bedzyk, 1997), similar to that observed in *E. coli* (Prince *et al.*, 1995).

Multicellular systems

Multicellular eukaryotic organisms such as transgenic insects, mammals and plants are capable of expressing engineered protein constitutively or in a tissue specific manner. Tissue-specific expression allows the accumulation of the recombinant protein in a selected compartment (targeting signals tagged on the proteins), or storage organ like a gland facilitating purification of the protein. Moreover, these systems may provide appropriate post-translational modifications and thus enable more appropriate folding of the most complex proteins.

Insects

Baculovirus-mediated insect cell protein expression systems have been developed for many years and are now commonly used for the expression of foreign proteins in different insect cell lines (O'Reilly *et al.*, 1992; see Hunt, 2005 for review). The availability of a strong polyhedrin gene promoter allows a high level of recombinant protein production in the baculovirus expression system. In addition, this system is capable of mediating a variety of common eukaryotic post-translational modifications (O'Reilly *et al.*, 1992). Though the most traditional and commonly used insect cell lines for exogenous protein production are the Sf9 and Sf21 cell lines, which are derived from *Spodoptera frugiperda* pupal ovarian tissue (Vaughn *et al.*, 1977), a *B. mori*-derived cell line (BmN insect cells) is also available. A recent study reports the use of a newly developed Bac-to-Bac/BmNPV (nuclear polyhedrosis virus) system (Motohashi *et al.*, 2005) to express a small MaSp 1 cDNA (*N. clavata*) into BmN insect cells and transfected silkworm larvaes (Zhang *et al.*, 2007; see subsection entitled 'Expression of native silk cDNAs' above). Unfortunately, the 70 kDa recombinant EGFP-MaSp 1 protein was highly insoluble and its yields were extremely low both in insect cells and in transfected larvae. The primary cause for this low yield was attributed to premature aggregation of the silk proteins (Zhang *et al.*, 2007).

The engineering of transposon-based gene vectors has allowed the production of several transgenic insects (for review see: Handler, 2001; Atkinson *et al.*, 2001; Handler, 2002; Atkinson, 2002; Robinson *et al.*, 2004). In Lepidoptera, the *piggyBac* transposable element was instrumental in the development of insect vector systems that were used to generate stable germ-line transgenic *B. mori* silkworms (Tamura *et al.*, 2000). Very soon after, the production of fibrous proteins was achieved in insect systems. For instance, a fusion cDNA coding partly for the *B. mori* fibroin L-chain and a partial human type III collagen (helical domain) was successfully expressed in transfected silkworms (Tomita *et al.*, 2003). The use of a fibroin L-chain specific promoter to control the fibroin–collagen sequence allowed targeted transgene expression into the silk glands of transgenic silkworms. Unfortunately, this fibroin–collagen fusion protein only represented 3.67% of the total protein extracted from cocoons (Tomita *et al.*, 2003).

Mammals

Production of recombinant proteins in mammalian cell cultures is feasible and this system has been extensively used for the production of most currently commercialized monoclonal antibodies (Pollock *et al.*, 1999). The use of mammary gland cell cultures allows secretion of the recombinant proteins. All types of mammalian cell culture systems require more delicate handling than simple organisms like bacteria and yeast. In addition, although they can be grown using steel tank bioreactors, large-scale bioreactor production is rather expensive. For instance, at a 10 000 L fermentation scale, the cost of purified antibodies is between 1000 and 2000 US$ per gram (Werner, 1998). Thus, other systems such as transgenic animal and plant systems are being investigated.

Because most cell culture systems are onerous, the use of transgenic animals, more specifically dairy animals (mammary gland expression system), to produce specific proteins, may be more adaptable for low-cost, large-scale production. In this system, the gene encoding the protein of interest is fused to milk specific gene regulatory elements (whey acid proteins, lactalbumin or casein genes) and the transgene is introduced by pronuclear microinjection into fertilized ovules collected from a female donor. The embryos are implanted in the uterus of a recipient female and are carried to term. The resulting offspring are screened and the animals containing the transgene are identified. Once mature, the transgenic animal can be hormonally induced to lactate for collection of the recombinant proteins or bred to generate a herd of transgenic animals for large-scale production (reviewed in Pollock *et al.*, 1999). Caprine mammals such as dairy goats have a gestation period far shorter than cattle (5 months versus 9 months) and, thus, are attractive for such large-scale production. Indeed, a doe has an average milk production of 600–800 L per 300 days of lactation. Human antibody expression levels of 14 mg/ml were observed for a female caprine founder versus 5 mg mL^{-1} for its offspring (Pollock *et al.*, 1999).

As with any system, there are problems associated with expression of recombinant proteins in transgenic animals, not the least of which being the ethical arguments against this approach. However, there are also technical hurdles in that protein extraction can be difficult, particularly from a high-protein, high-fat environment such as milk.

Plants

Of all the expression systems, plants require little material input and can produce significant levels of multimeric and fully processed recombinant proteins (Hondred *et al.*, 1999; Perrin *et al.*, 2000;). Transformation systems are available and the use of strong viral promoters such as the cauliflower mosaic virus CaMV 35S promoter leads to the constitutive expression of the transgene in essentially all plant tissues. Moreover, targeting of foreign protein accumulation to special organelles within the cell, such as chloroplasts or endoplasmic reticulum (ER), is possible by adding appropriate signals to the sequence of interest. As an example, plants such as tobacco that are relatively easy to transform and regenerate, were extensively and successfully used for the production of foreign proteins such as human collagen (Ruggiero *et al.*, 2000), or growth hormone (Staub *et al.*, 2000).

Plants have also been used for the production of antibodies or 'plantibodies'. These plant bioreactors are expected to generate 10 kg of antibodies per acre in tobacco, maize, soybean and alfafa (Larrick and Thomas, 2001). Some reports also show the expression of transgenic proteins in plant seeds of corn (Russell, 1999), soybean (Zeitlin *et al.*, 1998), tobacco (Fiedler and Conrad, 1995), and barley (Horvath *et al.*, 2000).

Chloroplast transformation is also an option in plant transgenics. As chloroplasts are not normally transmitted through pollen (Daniell, 2002), plastid transformation significantly reduces the risk of gene escape into neighboring wild populations. However, plastid transformation may not be the method of choice for highly repetitive sequences like silks. Since the introduction of transgenes relies on homologous recombination, genetic instability like that seen in bacterial systems may be problematic.

1.6 Artificial material production, properties and performance

We encourage the reader to explore other chapters in this book that are devoted to these subjects.

1.7 Conclusions

Through the continued exploration of the nature, process and production of protein-based biomaterials, a greater understanding of the structure/function relationships of structural protein materials is certain. Through the integration of this knowledge with bio-processing systems and recombinant DNA technologies, there is great promise for future production of bio-inspired materials with novel and desired properties for industrial and medical applications. We have made significant inroads into the understanding of the primary structure of these natural protein polymers and extensive studies have revealed much about their secondary and tertiary assemblies as well. With current recombinant engineering and cellular production systems, large quantities of pre-polymer subunits can be produced and purified. Coupling these living production systems with material spinning technologies will complete the process needed to prove the concept and provide the pipeline for large-scale production of bio-inspired materials. The next chapters in this book address these latter issues and demonstrate that the exciting future of bio-inspired materials is indeed on the horizon.

1.8 References

Andersen SO (1970), 'Amino acid composition of spider silk', *Comp Biochem Physiol*, **35**, 705–711.

Anderson JP, Cappello J, and Martin DC (1994), 'Morphology and primary crystal structure of a silk-like protein polymer synthesized by genetically engineered *Escherichia coli* bacteria', *Biopolymers*, **34**, 1049–1057.

Arcidiacono S, Mello C, Kaplan D, Cheley S, and Bayley H (1998), 'Purification and characterization of recombinant spider silk expressed in *Escherichia coli*', *Appl Microbiol Biotechnol*, **49**, 31–38.

Atkinson PW (2002), 'Genetic engineering in insects of agricultural importance', *Insect Biochem Mol Biol*, **32**, 1237–1242.

Atkinson PW, Pinkerton AC, O'Brochta DA (2001), 'Genetic transformation systems in insects', *Annu Rev Entomol*, **46**, 317–346.

Baneyx F (1999), 'Recombinant protein expression in *Escherichia coli*', *Curr Opin Biotechnol*, **10**, 411–421.

Beckwitt HR and Arcidiacono S (1994), 'Sequence conservation in the C-terminal region of spider silk proteins (spidroin) from *Nephila clavipes* (Tetragnathidae) and *Araneus bicentenarius* (Araneidae)', *J Biol Mol*, **9**, 6661–6683.

Bell EC and Gosline JM (1996), 'Mechanical design of mussel byssus: material yield enhances attachment strength', *J Exp Biol*, **199**, 1005–1017.

Capello J, Crissman J, Dorman M, Mikolajczak M, Textor G, Marquet M, and Ferrari F (1990), 'Genetic engineering of structural protein polymers', *Biotechnol Prog*, **6**, 198–202.

Cereghino JL and Cregg JM (2000), 'Heterologous protein expression in the methylotrophic yeast *Pichia pastoris*', *FEMS Microbiol Rev*, **24**, 45–66.

Cereghino GPL, Cereghino JL, Ilgen C, and Cregg JM (2002), 'Production of recombinant proteins in fermenter cultures of the yeast *Pichia pastoris*', *Curr Opin Biotechnol*, **13**, 329–332.

Coyne KJ, Qin XX, and Waite JH (1997), 'Extensible collagen in Mussel byssus: a natural block copolymer', *Science*, **277**, 1830–1832.

Cregg JM and Madden KR (1988), 'Development of the methylotrophic yeast *Pichia pastoris*, as a host system for the production of foreign proteins', *Dev Ind Microbiol*, **29**, 33–41.

Cunniff PM, Fossey SA, Auerbach A, Song JW, Kaplan D, Adams W, Eby D, and Vezie D (1994), 'Mechanical and thermal properties of dragline silk from *Nephila clavipes*', *Polym Adv Technol*, **5**, 401–410.

Daniell H (2002), 'Molecular strategies for gene containment in transgenic crops', *Nat Biotechnol*, **20**, 581–586.

Denny MW (1976), 'The physical properties of spider silks and their role in the design of orb webs', *J Exp Biology*, **65**, 483–505.

Denny MW (1980), 'Silks—their properties and functions' in '*The mechanical properties of biological materials'*, Cambridge, Cambridge University Press, 247–272.

Dicko C Vollrath F, Kenney JM (2004), 'Spider silk protein refolding is controlled by changing pH', *Biomacromolecules*, **5**, 704–710.

Dong Z, Lewis RV, and Middaugh CR (1991), 'Molecular mechanisms of spider silk elasticity', *Arch Biochem Biophys*, **284**, 53–57.

Fahnestock SR and Bedzyk LA (1997), 'Production of synthetic spider dragline silk protein in *Pichia pastoris*', *Appl Microbiol Biotechnol*, **47**, 33–39.

Fiedler U and Conrad U (1995), 'High-level production and long-term storage of engineered antibodies in transgenic tobacco seeds', *Biotechnol*, **13**, 1090–1093.

Fahnestock SR and Irwin SL (1997), 'Synthetic spider dragline silk proteins and their production in *Escherichia coli*', *Appl Microbiol Biotechnol*, **47**, 23–32.

Fossey SA, Nemethy G, Gibson KD, and Scheraga HA (1991), 'Conformational energy studies of beta-sheets of model silk fibroin peptides. I. Sheets of poly(Ala-Gly) chains', *Biopolymers*, **31**, 1529–1541.

Fukushima Y (1998), 'Genetically engineered syntheses of tandem repetitive polypeptides consisting of glycine-rich sequence of spider dragline silk', *Biopolymer*, **45**, 269–279.

Fukushima Y (2000), 'Secondary structural analysis in the solid state for analogous sequential polypeptides of glycine-rich sequence of spider dragline silk', *Polym Bull*, **45**, 237–244.

Gosline JM, Denny MW, and Demont ME (1984), 'Spider silk as a rubber', *Nature*, **309**, 551–552.

Gosline JM, Demont ME, and Denny MW (1986), 'The structure and properties of spider silks', *Endeavour*, **10**, 37–43.

Gosline JM, Guerette PA, Ortlepp CS, and Savage KN (1999), 'The mechanical design of spider silks: from fibroin sequence to mechanical function', *J Exp Biol*, **202**, 3295–3303.

Gosline J, Lillie M, Carrington E, Guerette P, Ortlepp C, and Savage K (2002), 'Elastic proteins: Biological roles and mechanical properties', *Phil Trans R Soc Lond B*, **357**, 121–132.

Grzelak K (1995), 'Control of expression of silk protein genes', *Comp Biochem Physiol B*, **110**, 671–681.

Guerette PA, Ginzinger DG, Weber BHF, and Gosline JM (1996), 'Silk properties determined by gland-specific expression of a spider fibroin gene family', *Science*, **272**, 112–115.

Handler AM (2001), 'A current perspective on insect gene transfer', *Insect Biochem Mol Biol*, **31**, 111–138.

Handler AM (2002), 'Prospects for using genetic transformation for improved SIT and new biocontrol methods', *Genetica*, **116**, 137–149.

Hannig G, and Makrides SC (1998), 'Strategies for optimizing heterologous protein expression in *Escherichia coli*', *Trends Biotechnol*, **16**, 54–60.

Hayashi CY and Lewis RV (1998), 'Evidence from flagelliform silk cDNA for the structural basis of elasticity and modular nature of spider silks', *Mol Biol*, **275**, 773–784.

Hayashi CY, Shipley NH, and Lewis RV (1999), 'Hypotheses that correlate the sequence, structure, and mechanical properties of spider silk proteins', *Int J Biol Macromol*, **24**, 271–275.

Hayashi CY and Lewis, RV (2000), 'Molecular architecture and evolution of a modular spider silk protein gene', *Science*, **287**, 1477–1479.

Hinman M and Lewis RV (1992), 'Isolation of a clone encoding a second dragline silk fibroin', *J Biol Chem*, **267**(27), 19320–19324.

Hirijidia DH, Do KG, Michal C, Wong S, Zax D, and Jelinski LW (1996), 'C_{13} NMR of *Nephila clavipes* major ampullate silk gland', *Biophys J*, **71**, 3442–3447.

Hochuli E (1988), 'Large-scale chromatography of recombinant proteins', *J Chromatogr*, **444**, 293–302.

Hochuli E (1990), 'Purification of recombinant proteins with metal chelate absorbent', *Genet Eng*, **12**, 87–98.

Hochuli E, Dobeli H, and Schacher A (1987), 'New metal chelate absorbent selective for proteins and peptides containing neighbouring histidine residues', *J Chromatogr*, **411**, 177–184.

Hondred D, Walker JM, Mathews DE, and Vierstra RD (1999), 'Use of ubiquitin fusions to augment protein expression in transgenic plants', *Plant Physiol*, **119**, 713–723.
Horvath H, Huang J, Wong O, Kohl E, Okita T, Kannagara LG, and Von Wettstein D (2000), 'The production of recombinant proteins in transgenic barley grains', *Proc Natl Acad Sci USA*, **97**, 1914–1919.
Hunt I (2005), 'From gene to protein: a review of new and enabling technologies for multi-parallel protein expression', *Prot Expr Purif*, **40**, 1–22.
Jelinski L (1998), 'Establishing the relationship between structure and mechanical function in silks', *Curr Opin Solid State Mat Sci*, **3**, 237–245.
Kuchar JA and Dooley DM (2001), 'Cloning, sequence analysis and characterization of the "Lysyl Oxidase" from *Pichia pastoris*', *J Inorg Biochem*, **83**, 193–204.
Kümmerlen J, Van Beek JD, Vollrath F, and Meier BH (1996), 'Local structure in spider dragline silk investigated by two-dimensional spin-diffusion nuclear magnetic resonance', *Macromol*, **29**, 2920–2928.
Larrick JW and Thomas DW (2001), 'Producing proteins in transgenic plants and animals', *Curr Opin Biotechnol*, **12**, 411–418.
Lazaris A, Arcidiacono S, Huang Y, Zhou J-F, Duguay F, Chretien N, Welsh EA, Soares JW, and Karatzas CN (2002), 'Spider silk fibers spun from soluble recombinant silk produced in mammalian cells', *Science*, **295**, 472–476.
Lewis RV (1992), 'Spider silk: the unraveling of a mystery', *Acc Chem Res*, **25**, 392–398.
Lewis RV, Hinman M, Kothakota S, and Fournier MJ (1996), 'Expression and purification of a spider silk protein: a new strategy for producing repetitive proteins', *Prot Expr Purif*, **7**, 400–406.
Lin CJ and Cregg JM (2001), 'Heterologous protein expression in the methylotrophic yeast *Pichia pastoris*', *FEMS Microbiol Rev*, **24**, 45–56.
Lombardi SJ, and Kaplan DL (1990), 'The amino acid composition of major ampullate gland silk (dragline) of *Nephila clavipes* (Araneae, Tetragnathidae)', *J Arachnol*, **18**, 297–306.
Lotz B and Cesari FC (1979), 'The chemical structure and the crystalline structures of *Bombyx mori* silk fibroin', *Biochimie*, **61**, 205–214.
Lotz B and Keith HD (1971), 'Crystal structure of poly-(L-Ala-Gly)II. A model for silk I', *J Mol Biol*, **61**, 201–215.
Lucas F (1964), 'Spiders and their silks', *Discovery*, 20–26.
Makrides SC (1996), 'Strategies for achieving high-level expression of genes in *Escherichia coli*', *Microbiol Rev*, **60**, 512–538.
Miao Y, Zhang Y, Nakagaki K, Zhao T, Zhao A, Meng Y, Nakagaki M, and Park EY (2006), 'Expression of spider flagelliform silk protein in *Bombyx mori* cell line by a novel Bac to-Bac/BmNPV baculovirus expression system', *Appl Microbiol Biotechnol*, **71**, 192–199.
Motohashi T, Shimojima T, Fukagawa T, Maenaka K, and Park EY (2005), 'Efficient large-scale protein production of larvae and pupae of silkworm by *Bombyx mori* nuclear polyhedrosis virus bacmid system', *Biochem Biophys Res Comm*, **326**, 564–569.
O'Reilly DR, Miller LK, and Luckow VA (1992), *Baculovirus expression vectors: A laboratory manual*, Oxford, Oxford University Press, 216–234.
Ouroudjev E, Soares J, Arcidiacono S, Thompson JB, Fossey SA, and Hansma HG (2002), 'Segmented nanofibers of spider dragline silk: atomic force microscopy and single-molecule force microscopy', *Proc Natl Acad Sci*, **99**(Supplement 2), 6460–6465.
Parkhe AD, Seeley SK, Gardner K, Thompson L, and Lewis RV (1997), 'Structural studies of spider silk proteins in the fiber', *J Mol Recog*, **10**, 1–6.
Perez-Rigueiro J, Viney C, Llorca J, Elices M (2000), 'Mechanical properties of single-brin silkworm silk', *J Appl Polym Sci*, **75**(10), 1270–1277.

Perrin Y, Vaquero C, Gerrard I, Sack M, Drossard J, Stoger E, Christou P, and Fischer R (2000), 'Transgenic pea seeds as bioreactors for the production of single chain Fv fragment (scFv) antibody used in cancer diagnosis and theraphy.' *Mol Breed,* **6**, 345–352.

Peters HM (1955), 'Über den spinnapparat von *Nephila madagascariensis* (Radnetzspinnen Argiopidae)', *Z Naturforsch*, **10**, 395–404.

Pollock DP, Kutzko J, Birck-Wilson E, Williams JL, Echelard Y, and Meade HM (1999), 'Transgenic milk as a method for the production of recombinant antibodies', *J Immunol Methods*, **231**, 147–157.

Prince JT, McGrath KP, DiGirolamo CM, and Kaplan DL (1995), 'Construction, cloning, and expression of synthetic genes encoding spider dragline silk', *Biochemistry,* **34**, 10879–10885.

Pujol JP (1970), 'Le collagène du byssus de *Mytilus edulis* L', *Z Zellforsch*, **104**, 358–374.

Qin XX, Coynes KJ, and Waite JH (1997), 'Tough tendons. Mussel byssus has collagen with silk-like domains', *J Biol Chem*, **272**(51), 32623–32627.

Qin XX and Waite JH (1998), 'A potential mediator of collagenous block copolymer gradients in mussel byssal threads', *Proc Natl Acad Sci USA*, **95**, 10517–10522.

Qu Y, Payne SC, Apkarian RP, and Conticello VP (2000), 'Self-assembly of a polypeptide multi-block copolymer modeled on dragline silk proteins', *J Am Chem Soc*, **122**, 5014–5015.

Robinson AS, Franz G, and Atkinson PW (2004), Insect transgenesis and its potential role in agriculture and human health', *Insect Biochem Mol Biol*, **34**, 113–120.

Ruggiero F, Exposito JY, Bournat P, Gruber V, Perret S, Comte J, Olagniere B, Garrone R, and Thiesen M (2000), 'Triple-helix assembly and processing of human collagen produced in transgenic tobacco plants', *FEBS*, **23401**, 132–136.

Russell DA (1999), 'Feasibility of antibody production in plants for human therapeutic use', *Curr Opin Biotechnol*, **10**, 382–386.

Scheller J, Guhrs K-H, Grosse F, and Conrad U (2001), Production of spider silk proteins in tobacco and potato. *Nature Biotechnol*, **19**, 573–577

Sehnal F and Akai H (1990), 'Insect silk glands: their types, development and function, and effects of environmental factors and morphogenetic hormones on them', *Int J Insect Morphol Embryol*, **19**, 79–132.

Simmons A, Michal C, and Jelinski L (1996), 'Molecular orientation and two component nature of the crystalline fraction of spider dragline silk', *Science*, **271**, 84–87.

Staub JM, Garcia B, Graves J, Hajdukiewicz PT, Hunter P, Nehra N, Paradkar V, Schlittler M, Carroll JA, Spatola L *et al.* (2000), 'High-yield production of a human therapeutic protein in tobacco chloroplasts', *Nat Biotechnol*, **18**, 330–338.

Stauffer S, Coguill S, Lewis R (1994), 'Comparison of physical properties of three silks from *Nephila clavipes* and *Araneus gemmoides*', *J Arachnol*, **22**, 5–11.

Tamura T, Thibert T, Royer C, Kanda T, Eappen A, Kamba M, Kômoto N, Thomas J-L, Mauchamp B, Chavancy G, Shirk P, Fraser M, Prud'homme J-C, and Couble P (2000), 'A *piggyBac* element-derived vector efficiently promotes germ-line transformation in the silkworm *Bombyx mori* L', *Nat Biotechnol*, **18**, 81–84.

Tanaka K, Inoue S, and Mizuno S (1999a), 'Hydrophobic interactions of P25, containing Asn-linked oligosaccharide chains, with the H-L complex of silk fibroin produced by *Bombyx mori*', *Insect Biochem Mol Biol*, **29**, 269–276.

Tanaka K, Kajiyama N, Ishikura K, Waga S, Kikuchi A, Ohtomo K, Takagi T, and Mizuno S (1999b), 'Determination of the site of disulfide linkage between heavy and light chains of silk fibroin produced by *Bombyx mori*', *Biochem Biophys Acta*, **1432**(1), 92–103.

Tatham AS and Shewry PR (2000), 'Elastomeric proteins: biological roles, structures and mechanisms', *Trends Biol Sci*, **299**, 567–571.

Teulé F, Furin WA, Cooper AR, Duncan JR, Lewis RV (2007), 'Modifications of spider silk sequences in an attempt to control the mechanical properties of the synthetic fibers', *J Mater Sci*, **42**, 8974–8985.

Thiel BL, Kunkel DD, and Viney C (1994), 'Physical and chemical microstructure of spider dragline silk: a study by analytical transmission electron microscopy', *Biopolymers*, **34**, 1089–1097.

Tomita M, Munetsuna H, Sato T, Adachi T, Hino R, Hayashi M, Shimizu K, Nakamura N, Tamura T, and Yoshizato K (2003), 'Transgenic silkworms produce recombinant human type III procollagen in cocoons', *Nat Biotechnol*, **21**, 52–56.

Vaughn JL, Goodwin GJ, Tompkins P, and McCawley P (1977), 'The establishment of two cell lines from the insect *Spodoptera frugiperda*', *In Vitro*, **13**, 213–217.

Waite JH (1992), 'The formation of mussel byssus: anatomy of a natural manufacturing process', *Results Prob Cell Differ*, **19**, 27–54.

Waite JH, Vaccaro E, Sun C, and Lucas JM (2002), 'Elastomeric gradients: a hedge against stress concentration in marine holdfasts?', *Phil Trans R Soc Lond B*, **357**, 143–153.

Warsame A, Vad R, Kristensen T, and Beate Øyen T (2001), 'Characterization of a gene encoding a *Pichia pastoris* protein disulfide isomerase', *Biochem Biophys Res Comm*, **281**, 1176–1182.

Werner RG (1998), 'Innovative and economic potential of mammalian cell culture', *Arzneim Forsch Drug Res*, **48**, 423.

Winkler S, Szela S, Avtges P, Valluzzi R, Kirschner DA, and Kaplan D (1999), 'Designing recombinant spider silk proteins to control assembly', *Int J Biol Macrom*, **24**, 265–270.

Winkler S, Wilson D, and Kaplan D (2000), 'Controlling β-sheet assembly in genetically engineered silk by enzymatic phosphorylation/dephosphorylation', *Biochem*, **39**, 12739–12746.

Work RW and Young CT (1987), 'The amino acid compositions of major and minor ampullate silks of certain orb-web-building spiders (Araneae, Araneidae)', *J Arachnol*, **15**, 65–80.

Xu M, and Lewis RV (1990), 'Structure of a protein superfiber: spider dragline silk', *Proc Natl Acad Sci USA*, **87**, 7120–7124.

Zeitlin L, Olmsted SS, Moench TR, Co MS, Martinell BJ, Paradkar VM, Russell DR, Queen C, Cone RA, and Whaley KJ (1998), 'A humanized monoclonal antibody produced in transgenic plants for immunoprotection of the vagina against genital herpes', *Nat Biotechnol*, **16**, 1361–1364.

Zhang Y, Hu J, Miao Y, Zhao A, Zhao T, Wu D, Liang L, Miikura A, Shiomi K, Kajiura Z, and Nakagaki M (2007), 'Expression of EGFP-spider dragline silk fusion protein in BmN cells and larvae of silkworm showed the solubility is primary limit for dragline protein yield', *Mol Biol Rep*, online publication, 10.1007/s11033-007-9090-6.

Zhou C-ZF, Confalonieri N, Medina Y, Zivanovic Y, Esnault C, Yang T, Jacquet M, Janin J, Duguet M, and Perasso R (2000), 'Fine organization of *Bombyx mori* heavy chain silk gene', *Nucleic Acids Res*, **28**, 2413–2419.

Zhou Y, Wu S, and Conticello V (2001), 'Genetically directed synthesis and spectroscopic analysis of a protein polymer derived from a Flagelliform silk sequence', *Biomacromolecules*, **2**(1), 111–125.

2
Purification of protein solutions

S. HARCUM
Clemson University, USA

Abstract: Purification and separation methods currently used for proteins are discussed, with a special emphasis on the application of these techniques to biologically inspired textile proteins. Many natural proteins found in a cell are enzymes that only function when soluble; in contrast, recombinant DNA proteins destined to be textiles will be required to function as insoluble or solid materials. The methods, which have to separate the target protein and purify it without severe losses of material, include filtration and centrifugation for insoluble removal, ultrasonic shock for mechanical cell disruption, chromatographic methods for soluble separations, and drying and crystallization for finishing.

Key words: protein purification, chromatography, recombinant DNA, textile proteins./

2.1 Introduction

Synthetic proteins, with a design founded in biological inspiration, can find applications in textile materials. These 'biologically inspired textile proteins', derived by recombinant DNA techniques, have the potential for precisely tailored characteristics, such as strength, elasticity, solubility, hydrophobicity, size and charge. The variety of characteristics available to biologically inspired textile proteins are due to the ability to determine the amino acid sequence within a protein at the DNA level, and then produce the biologically inspired textile protein in a rapidly growing organism under controlled environmental conditions. Once produced, however, the biologically inspired textile protein will require processing steps to separate it from the proteins the host organism cellular components and purify it before spinning or other fiber formation steps.

Many natural proteins found in a cell are enzymes that only function when soluble; in contrast, recombinant proteins destined to be textiles will be required to function as insoluble or solid materials. Yet, all proteins begin as soluble proteins inside a cell. Therefore, one of the greatest challenges for biologically inspired textile proteins will be the separation and purification of the target protein without severe losses of material, i.e. low yields, due to premature aggregation.

Therapeutic recombinant proteins, such as recombinant human insulin, are very expensive compared to textiles, on the order of \$10 000 per kilogram for 'inexpensive' therapeutic proteins. Fortunately for the healthcare consumer, the amount of recombinant protein needed (a dose) is very small (milligram amounts). Additionally, people are willing to pay significant sums of money for drugs to sustain and improve quality of life. Since biologically inspired textile proteins are also produced by recombinant DNA techniques, the tools needed to separate and purify these proteins will need to rely on the methods developed and refined by the biotechnology industry, where the focus of the traditional biotechnology industry is to obtain a highly purified protein with less regard to costs and yields than will be required or desired for biologically inspired textile proteins.

Recombinant proteins are expressed by the host organism and can remain inside the cell or be secreted by the host, termed intracellular and extracellular, respectively. From a protein purification perspective, the fate of the recombinant protein, intracellular or extracellular, shapes the initial separation and purification steps. Proteins expressed extracellularly are often much easier and cheaper to purify, since steps are not required to separate the desired protein from most of the host cellular components, including host proteins. A host organism can contain up to 1000 different protein species, DNA, RNA, lipids, and polysaccharides. Intracellular proteins are the source for most therapeutic proteins due to the rapid growth characteristics of the bacteria *Escherichia coli*. This rapid growth rate allows for large amounts of protein to be manufactured in a short time; however, the overall protein yield for a typical intracellular therapeutic protein is 5 to 20%, compared with over 50% for extracellular proteins. Additionally, for most therapeutic proteins and many industrial enzymes (including non-recombinant) the separation and purification steps account for over 50% of the total manufacturing costs.

Protein separation and purification methods are loosely grouped by the principle of separation, i.e. size, solubility, or charge. A series of separation and purification methods are used that rely on different principles of separation (termed orthogonal methods) in order to obtain a purified target protein. In general, protein separation methods have one of four basic functions: (1) removal of insoluble material; (2) primary isolation and/or concentration; (3) removal of major contaminants; and (4) final product preparation. Since water is the major contaminant that dilutes the product, process economics usually position the water removal steps early in the purification process. Many of the equipment units used for isolation and contaminant removal are the same, thus the distinction between these steps is often not made. Figure 2.1 shows two examples of flow diagrams outlining the major steps used to purify extracellular and intracellular proteins for therapeutic proteins, which are often injectable drugs with very stringent purity standards. The order of the chromatography steps depends on the characteristics of the target protein and contaminants. For biologically inspired textile proteins, there may be a limit on water removal that is much lower than that allowed for therapeutic proteins, owing to the propensity of textile proteins to precipitate (become a solid) and the difficulties one encounters with re-solubilization, as has been observed with

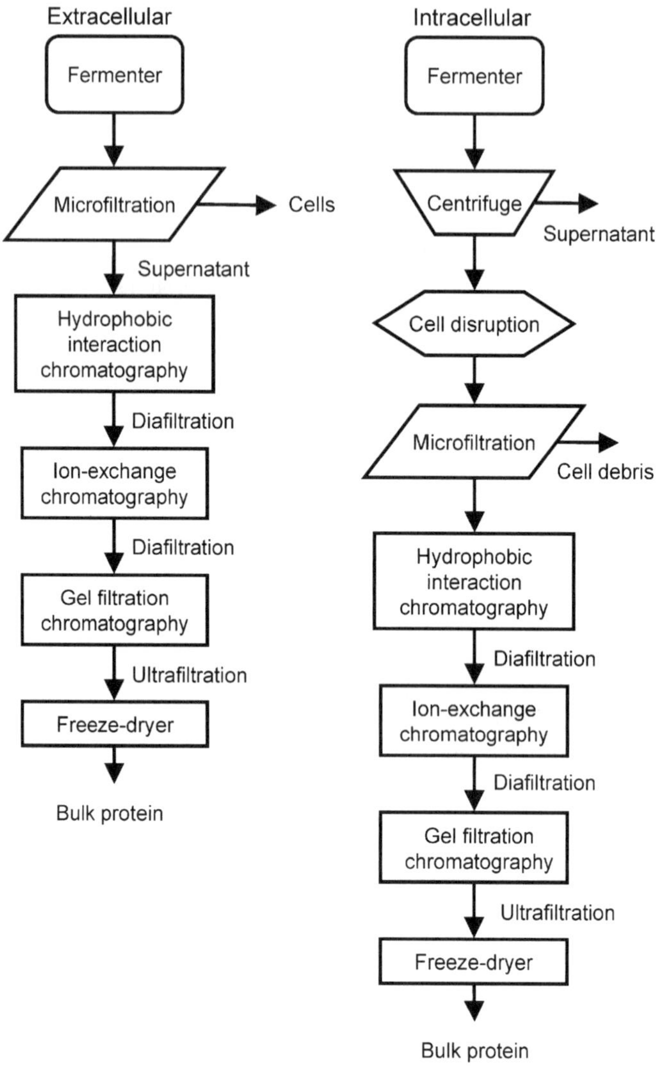

2.1 Simplified diagram of the major separation and purification steps for extracellular and intracellular recombinant proteins.

natural silk proteins. Additionally, it is anticipated that the number of steps will be lower for biologically inspired textile proteins compared with therapeutic proteins.

2.2 Insoluble removal

Independent of the location of the desired protein (extracellular versus intracellular), the first step after the fermentation step is separation of the cells and culture broth, an isolation step. If the protein is extracellular, the cells are discarded.

Conversely, if the protein is intracellular, the culture broth is discarded. There are two well-established methods to separate soluble and insoluble materials. One method is based on size differences (filtration) and the second method is based on density differences (centrifugation). A third method, coagulation and/or flocculation, relies on the addition of an agent that causes the cells or other soluble materials to aggregate or precipitate, which then is followed by a size- or density-based separation step. These three methods will be described in more detail below with information regarding the application of the method to biologically inspired textile proteins.

2.2.1 Filtration

The purpose of standard filtration is to remove particles from a solution, where the principle of separation is the size differences between the particles and the fluid. Standard filtration is routinely used for extracellular fermentation products, including antibiotics and industrial reagents, such as citric acid. For recombinant protein fermentations, standard filtration is only used for extracellular proteins. In standard filtration, two fluid flow patterns are commonly used: conventional and crossflow, as shown in Fig. 2.2. For conventional filtration, the flow of the feed, which can be the harvested broth from the ferementer, is perpendicular to the filter media. The permeate stream is the fluid that passes through the filter. In conventional filtration, filter aids are often added to the feed to prevent cake compression. A common filter aid is diatomaceous earth. The addition of a filter aid usually precludes the further use of the cells. For crossflow filtration the flow of the feed is tangential to the filter media, the cells exit the system in the retentate as a concentrated cell slurry, and the permeate exits as clarified culture broth. Since it can be difficult to maintain sterility of the cells deposited on a filter media, if it is desired to save the cells, tangential flow is more desirable. Depending on the amount of material to be processed, continuous filtration versus batch filtration devices can be used for both conventional and crossflow filtration. For biologically inspired textile proteins, an extracellular product is more desirable, as a filtration step to remove cells is relatively inexpensive and results in fewer required purification steps.

2.2.2 Centrifugation

Centrifuges separate solids from liquids based on the density difference between the solid particles and fluid. The solid particles are typically whole cells or cell debris. Centrifuges can be operated in batch or continuous modes. The cells or culture broth can be easily saved for further purification steps. Owing to the small difference in density between cells (<1.1 g cm^{-3}) and the culture broth (1.0 g cm^{-3}), relatively high rotational speeds are used to decrease the equipment operation times, i.e. for a centrifuge spinning at 1000 rpm that requires 5 h to completely

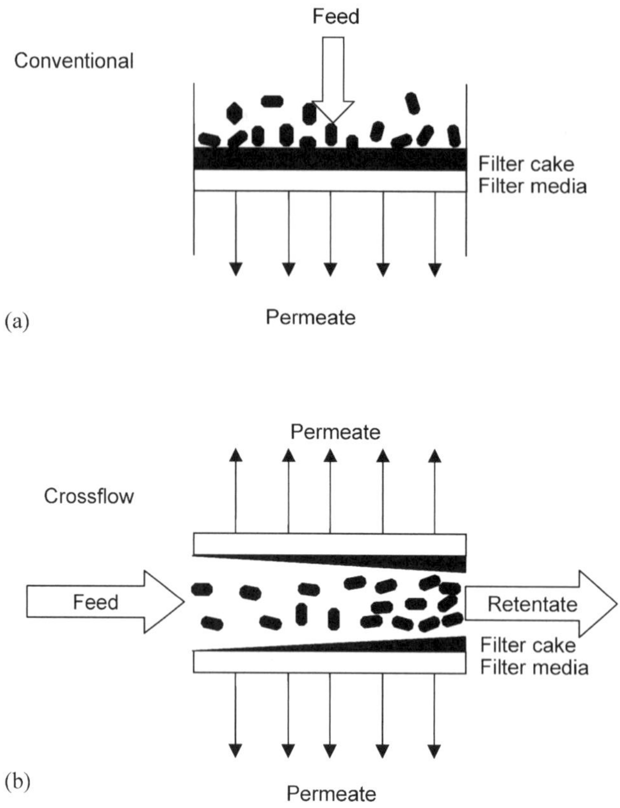

2.2 Schematic diagram of (a) conventional and (b) crossflow filtration.

pellet the cells, the same centrifuge spun at 5000 rpm would only require 12 min. The g-force of a centrifuge depends on the rotational speed and the radius of the centrifuge, where the g-force is proportional to both parameters. Cell debris requires significantly high rotational speeds, often approaching 10 000 *g*. Centrifuges are more expensive to purchase than filtration devices, while operational costs are usually lower. Since centrifuges are closed, retaining the cell pellet or culture broth for further processing is possible. Thus, for intracellular biologically inspired textile proteins, centrifugation or crossflow filtration may be appropriate.

2.2.3 Coagulation/flocculation

Coagulation and flocculation are used in conjunction with filtration and centrifugation to improve the efficiency of these other processes. For example, if the cells to be removed from a culture broth are very small, it is possible that the filter media can become fouled (or clogged) owing to the cells filling the void space in the filter media. However, if the cells can be made to aggregate into larger more rigid particles, filtration becomes more efficient because of the lower pressure drops required across the filter media. The same principle can be applied to centrifugation, an agent can be added to the culture broth that causes the cells to aggregate with this agent, such that the cell aggregate has a higher density. Centrifugation efficiencies increase proportionally based on the density difference. Coagulation and flocculation agents are normally relatively inexpensive; however, the fate of the cells may preclude some agents. Typically, diatomaceous earth and calcium chloride can be used as flocculation agents. High-molecular-weight, water-soluble organic compounds can also be used; however, waste disposal costs may be higher. Thus, for extracellular biologically inspired textile proteins, coagulation and flocculation may be appropriate alternatives, since the cells are normally discarded.

2.3 Cell disruption

The purpose of cell disruption is to release the product from the cells for intracellular products, such as a recombinant protein. *E. coli* does not secrete recombinant proteins, so recombinant proteins made in this host cell will require a cell disruption step to release the protein into the soluble fraction, such as a biologically inspired textile protein. Yeast strains can be engineered to either secrete or retain a recombinant protein, thus a cell disruption step may be needed. Mammalian cells usually secrete the recombinant protein, such that the cells are usually removed, and not disrupted. Additionally, mammalian cells do not have rigid cell walls like yeast and bacteria, and thus can be disrupted by far gentler treatments, if only intracellular expression can be obtained.

Cell disruption causes nearly all the contents of a cell to enter the soluble fraction along with the desired protein. The isolation and purification steps then must remove host cell proteins, DNA, RNA, lipids, and polysaccharides. Cell disruption methods can be divided into two types: chemical and mechanical. Industrially, mechanical cell disruption methods are preferred because of the lower operational costs and because there is no need to add chemicals or materials to the process that need to be removed at a later stage.

2.3.1 Mechanical

Mechanical cell disruption methods rely on pressure differences between the outside and the inside of the cells to break open the cells by physically moving the fluid around the cells or by moving the cell relative to itself (shearing the cell). The pressure differential can be generated by cavitation (bubble burst), high pressure on the fluid, and grinding (high pressure on cell). High physical pressure on the fluid or cells is more commonly used industrially owing to costs.

Ultrasonic shock

Ultrasonic shock or sonication is a commonly used laboratory-scale method to disrupt cells. This method uses high frequency sound waves to cause cavitations within the fluid. When a cavitation bubble bursts near a cell, a pressure gradient is generated that disrupts cell. Sonication is very expensive to use because of its low efficiency to generate cavitation, and the subsequent amount of heat generated due to this inefficiency. For heat-susceptible proteins, high levels of cooling are required, which further increases costs. At the laboratory scale (0.5 to 1.5 mL), sonication provides a rapid method of cell disruption. This method is fairly harsh and difficult to scale-up due to the heat removal requirements.

Pressure difference

Pressure difference methods use high pressures (20 to 40 kpsi) to break cells. For laboratory-scale devices (i.e. French presses) the cell slurry is pressurized, then forced through a small orifice with atmospheric pressure on the other side, where the sudden change in pressure causes the cells to burst. Industrial-scale devices pressurize the orifice chamber then the cell slurry is pumped into the chamber and forced out through the orifice to atmospheric pressure. Continous flow devices have a more even pressure drop across the entire run compared with batch devices, since the flow chamber is pressurized to a constant value. The cost of these devices is moderate and the process is fairly harsh. Cooling is required to counterbalance the friction effects of the culture broth moving through the small orifice; however, this cooling load per liter of fluid is significantly lower than sonication. Pressure difference cell disruption may be an appropriate method for intracellular biologically inspired textile proteins.

Grinding

There a number of grinding cell disruption methods, including ball mills, homogenizers, and grinders. All of these grinding methods use a high-speed impeller, glass beads, steel balls, or even flat plates to smash the cells and create shear stresses across the cell. Grinding units vary from laboratory- to industrial-scale. If a bead or ball system is used, the beads and balls must be separated from the disrupted cell debris slurry or cheap enough to discard with the cell debris after a second removal of insolubles step. These grinding methods can be batch or continuous and tend to be moderately harsh. Also, these methods are typically less expensive than pressure difference methods. Grinding methods for cell disruption may be appropriate for intracellular biologically inspired textile proteins.

2.3.2 Chemical (non-mechanical)

Osmotic shock

Osmotic shock relies on the chemical balance that cells maintain with their environment. Specifically, the osmotic balance cells have with respect to salts and sugars. Normally, cells have higher concentrations of salts and sugars inside the cell than in the culture broth, which causes an osmotic imbalance that is counterbalanced by the cell wall strength. Cells expend energy to bring many small molecules into the cell to maintain their osmotic balance. Also, as part of this osmotic balance, water freely diffuses across the membrane from the high water concentration outside the cell to the lower water concentration inside the cells. Thus, when cells are placed in a solution that has a lower concentration than normal, the net flow of water increases towards the inside of the cell. Eventually, the volume of a cell exceeds the cell wall strength capacity or cell membrane and the cells ruptures. This cell disruption method is inexpensive and gentle. For some very robust cells, osmotic shock is not sufficient to release the cell contents in a timely manner, so combining osmotic shock with freezing, enzyme degradation, or pressure disruption can enhance recoveries. For intracellular biologically inspired textile proteins produced in yeast, osmotic shock alone would not be sufficient to break open the cells.

Enzymes

Enzymes, such as lysozyme, can be used to digest the cell wall of most bacteria. Once the cell wall is digested, the cells burst owing to a weakened cell wall that cannot counterbalance the normal osmotic pressure of the cell. Other enzymes with or without lysozyme can be used to disrupt some bacterial species and yeast in order to enhance yield and/or decrease process time. Freezing and thawing cells treated with enzymes has also been used to improve efficiencies. Two major drawbacks for this method include high costs and the need to remove any enzymes added in later purification steps. For biologically inspired textile proteins, the addition of an enzyme may not be as problematic as it is for therapeutic proteins; however, the effect of the enzyme on the final material properties needs to be assessed to determine if enzymes are appropriate.

Solvent and detergents

Solvent and detergents are more often used with yeast and mammalian cells. Mammalian cells do not have cell wall, but have bilipid cell membrane, where solvents, such as toluene, can be used to dissolve the cell membrane, thus completely disrupting the cell. Acetone is a solvent that has been used to successfully disrupt yeast cells. Detergents that target cell membranes are mainly used for mammalian cell disruption. With solvent methods, the solvent is most often removed by evaporation. Removal of the detergent from the solution can be achieved by most chromatography steps. Solvent and detergent methods are less expensive than enzyme methods; however, they can increase the duration or number of later process steps. The effect of the solvent or detergent addition and need for removal needs to be assessed with respect to the final material properties of the biologically inspired textile protein.

2.4 Soluble protein separations

Once the majority of the insoluble particles have been removed, the target protein needs to be separated from other soluble host cell materials, such as host DNA, RNA, lipids, polysaccharides, and proteins. The degree of purification is determined by the end-use of the protein. For example, therapeutic proteins have extremely stringent purity standards, whereas industrial enzymes used in laundry detergent have much lower purity requirements. Biologically inspired textile proteins are more likely to have purity standards similar to industrial enzymes; however, material properties may be affected by contaminants necessitating a higher level of purity.

Separation of DNA and RNA from a target protein is usually the easiest separation owing to the highly negative nature of DNA and RNA under physiological conditions. However, this separation can be more challenging if the target protein is also highly negatively charged. Lipids are highly polar species, such that extraction and some chromatography steps can be used to remove this contaminant. Polysaccharides are most often removed by a de-pyrogen chromatography step. The remainder of this section will highlight the principles of the most commonly used separation and purification methods, starting with the least expensive method, extraction, and progressing to the most expensive method, chromatography.

2.4.1 Liquid–liquid extraction

Liquid–liquid extraction is a method of separating molecules based on the solubility of the species present in two immiscible phases, normally both liquids, one usually an organic solvent. The principle of separation relies on the target molecule having different solubility in the two phases, and different solubilities than the contaminants. Thus, the target molecule will be preferential transferred from one solvent to the other solvent with different affinities than the contaminants. Extraction is an equilibrium process, where the partitioning between the phases is defined as K, the partition coefficient, for each species in the system, as shown in Equation 2.1.

$$K = \frac{y}{x} \qquad [2.1]$$

where y and x represent the concentration of the molecule (solute) in the extraction solvent and waste solvent (raffinate), respectively. K is highly dependent on pH, temperature, solvents, and concentration of salts in the solvent phases. The sensitivity of K to process parameters has allowed for ‘tuning’ of the extraction

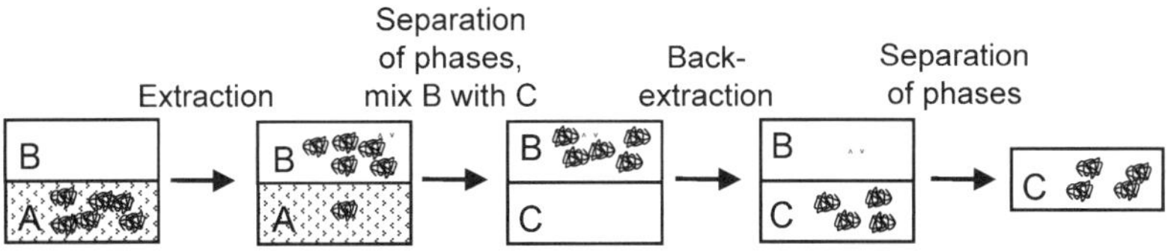

2.3 Schematic diagram of an extraction process. Solution B is contacted with Solution A, which contains all the target protein under conditions where the target protein has a greater affinity for Solution B. Solution A is removed and Solution B is contacted with Solution C, which may be the same solvent as was in Solution A. If so, the conditions are different from the first step (i.e. different pH, temperature, or salt concentration). The target protein transfers to Solution C and Solution B is removed.

process to favor the target molecule relative to the contaminant molecules. For example, penicillin was discovered to be highly soluble in an organic solvent at low pH and then almost completely insoluble in the same solvent at high pH. This allowed for the development of an inexpensive penicillin process, where penicillin is first extracted into the solvent at low pH, and then back-extracted into water at high pH. Figure 2.3 shows the basic steps in the extraction process, including the back-extraction step.

2.4.2 Aqueous two-phase extraction

Many proteins are irreversibly denatured by organic solvents, such that traditional liquid–liquid extraction using organic solvent was limited. A second-generation extraction system was developed where the solvents are aqueous, termed aqueous two-phase extraction. This newer extraction method relies on two immiscible solvents that are both aqueous, where one or both of the solvents contain a polymer that alters the solvent's density and solubility characteristics. Poly(ethylene)glycol (PEG)–dextran systems are commonly used to separate such proteins from host nucleic acids (DNA and RNA species) and polysaccharides. By changing the ratio of PEG to dextran, the K value for a particular target molecule can be varied to optimize the separation. The steps shown in Fig. 2.3 also represent the steps used for aqueous two-phase extraction. Owing to the aqueous environment, this method has been used to isolate and purify proteins with little denaturing of the protein. The concentration of the protein in the phases is still relatively low, so this method might be well-suited to biologically inspired textile protein to prevent self-assembly induced by concentration.

2.4.3 Precipitation

Precipitation or 'salting-out' is a relatively inexpensive way to concentrate and purify molecules, including proteins. A concentrated salt solution is slowly added to a solution of the target solute, the precipitates being removed periodically by filtration or centrifugation. Usually ammonium sulfate is used as the salt. Proteins assume structures in aqueous solutions that minimize the contact of the hydrophobic amino acid residues with the solvent (water), and maximize the contact between the polar amino acid residues and the solvent. Many protein characteristics, such as size, shape, and charge together determine the salt concentration necessary to precipitate a protein. Additionally, environmental factors, such as solvent and temperature, can be used to alter the salt concentration necessary for precipitation. The 'salting-out' concentration for any one protein needs to be determined experimentally. For proteins that are difficult to solublize or re-solubilize, such as many biologically inspired textile proteins, any precipitation should be avoided until contaminants have been removed to sufficient levels, or only used to remove the contaminants by precipitation.

2.4.4 Adsorption

Adsorption, like chromatography, relies on the different affinities solutes have for a solid surface. This is an equilibrium process (Equation 2.2) and can be described by an equilibrium constant (K_{eq}) that is dependent on the concentration of the solute [C], the concentration of empty adsorption sites [S], and the concentration of occupied adsorption sites [CS], as shown in Equation 2.3.

$$\mathrm{C} + \mathrm{S} \xleftrightarrow{K_{eq}} \mathrm{CS} \qquad [2.2]$$

$$K_{eq} = \frac{[\mathrm{CS}]}{[\mathrm{C}][\mathrm{S}]} \qquad [2.3]$$

Since [S] can not be readily measured during the process, various mechanisms have been proposed to describe the process mathematically in terms of measurable parameters. The most commonly used method to describe adsorption for proteins is the Langmuir isotherm method (Equation 2.4). An alternative description is the Freundlich isotherm, which has been used successfully to describe steroids, antibiotics, and hormones adsorption. In the Langmuir isotherm, the total number of adsorption sites S_{total} and the solute concentration determine the amount of solute that can be adsorbed (i.e. recovered).

$$[\mathrm{CS}] = \frac{K_{eq} S_{total} [\mathrm{C}]}{1 + K_{eq}[\mathrm{C}]} \qquad [2.4]$$

Each species in the system has an equilibrium constant. Additionally, species may even have a different number of total sites due to size constraints. The Langmuir equation predicts a saturation value of $K_{eq} S_{total}$[C] for [CS], if K_{eq} is small relative to 1. In contrast, the Langmuir equation predicts that all the sites will be occupied if K_{eq} is large for a particular species. Since adsorption is relatively inexpensive, this may be an appropriate separation and purification method for biologically inspired textile proteins. However, as the high concentration of the biologically inspired textile protein at the surface sites may be a problem, this method may prove to be better for removal of contaminants, rather than the adsorption of the biologically inspired textile protein.

2.4.5 Microfiltration and ultrafiltration

Microfiltration and ultrafiltration operate under the same principle of separation as standard filtration (size); however, the operational characteristics are different due to the very small pore sizes of the membranes used as the filter media. Microfiltration is defined to separate in the range 0.5 to 5 μm, which includes viruses, bacteria, and paint pigments. Ultrafiltration is defined to separate in the range 0.001 to 0.2 μm. Normally, ultrafiltration membranes are specified by the molecular weight that does not pass through the membrane, also called the cut-off molecular weight (i.e. a 10 000 cut-off means that most molecules smaller than 10 kDa will pass through the membrane and most molecules larger than 10 kDa will be retained by the membrane. Thus, ultrafiltration can be use to concentrate protein solutions in a continuous system using a tangential (crossflow) flow configuration.

In ultrafiltration processes a phenomenon called gelling can occur at the membrane due to the high concentration of protein that can develop in the boundary layer just above the membrane. Tangential flow and mixing can help reduce this phenomenon; however, a boundary layer always develops above the membrane. With regard to biologically inspired textile proteins, operation characteristics must minimize this phenomenon in order to prevent self-assembly under the local high concentration at the membrane.

2.4.6 Dialysis

Dialysis is a commonly used laboratory-scale process to remove salt, or reduce the salt concentration, from a solution. A semi-permeable membrane is used to contain the target protein. The target protein in solution is placed into the dialysis tubing and the dialysis tubing is sealed and placed into a large container of water, as shown in Fig. 2.4. The salt in the dialysis tubing moves from high concentration to low concentration via passive diffusion. The final salt concentration in the protein solution is the weighted average of the salt concentration based on the volumes of the starting protein solution and the water. As the volume of the solution in the dialysis tubing decreases, the protein concentration increases. Continuous dialysis is called diafiltration, and the protein solution is diluted with large amounts of water and then passed through an ultrafiltration device to remove both the salts and water. Both diafiltration and dialysis methods can be repeated by adding or contacting, respectively, the target solution with more water.

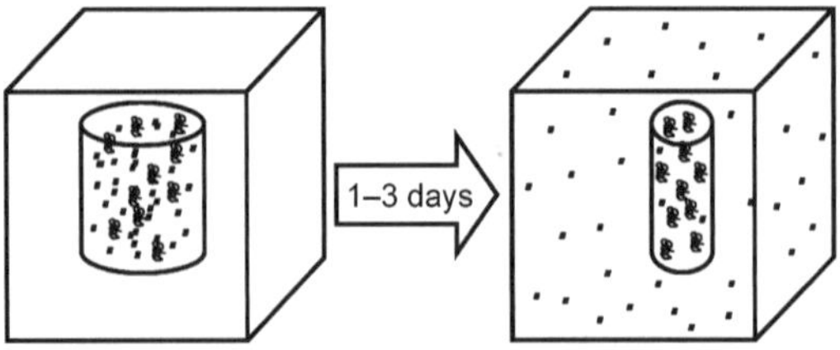

2.4 Schematic diagram for a dialysis system. The proteins (target and other large molecules) are represented by the globular shapes and the smaller molecular species are shown as dots. Initially, all the salt is in the dialysis tubing. After 1–3 days the salt and water exchange through the semi-permeable membrane reaching equilibrium. The salt concentration inside the dialysis tubing and outside the dialysis tubing are equal.

2.4.7 Chromatography

Chromatography is an adsorption process where the solid phase is stationary and packed into a column. The solutes are in the mobile phase and are passed through the column under pressure. The pressure drop in the column due to the frictional effects of flow through a packed bed require that the resins used are sufficiently stabilized to withstand these pressures without collapsing. Differences in affinity for the solid phase of the target protein and contaminants enable separation and purification of the target from other molecules. There are many types of solid surfaces that can be used as the stationary phase including silica, dextran, polyacrylamide, and agarose. Figure 2.5 shows the basic steps of chromatography. The target protein is collected in a fraction during the elute step. Depending on the protein and contaminants, there may be more than one wash, regeneration, and/or equilibrium steps required. Elution chromatography uses buffers with gradient properties in order to improve separations. Example elution profiles for three species are shown in Fig. 2.6.

Adsorption

Adsorption chromatography only differs from adsorption in how the process operates, not the principle of separation. For adsorption chromatography, the adsorbent is used as the stationary phase. The solute binds to the adsorbent via van der Waal forces and steric interactions. Since the adsorption sites are typical only on the outer surface of the stationary phase, fairly small particles are used as the stationary phase. Smaller-particle stationary-phase materials have higher frictional effects and, thus, larger pressure drops during operation. Thus, the stationary phase for adsorption chromatography needs to be able to handle the pressure drop necessary for the mobile phase to flow through the packed-bed column. Typical adsorbents include alumina- or silica-based resins, which are very rigid.

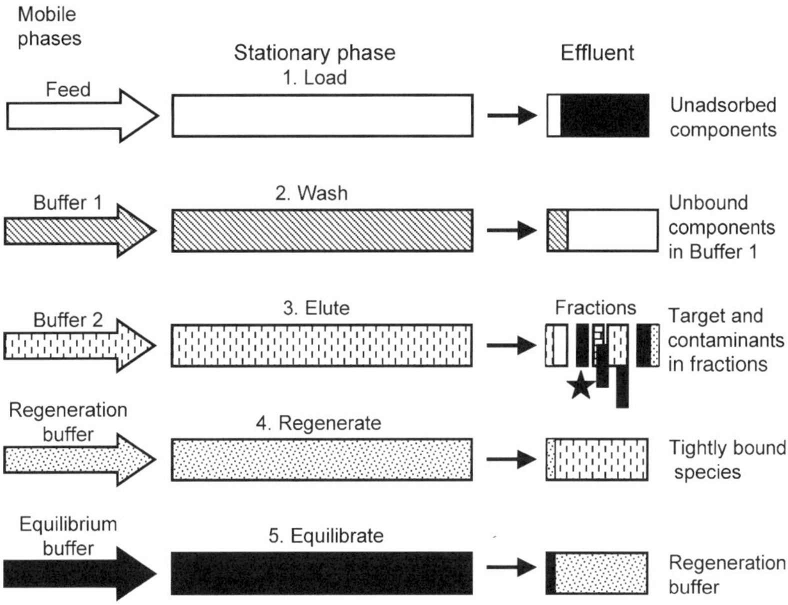

2.5 Schematic of the basic steps of chromatography.

Liquid–liquid partitioning

Liquid–liquid partitioning chromatography relies on differences in solubility of solutes in a liquid phase that is adsorbed onto the stationary phase. Most often the adsorbed liquid phase is a non-polar solvent, where the mobile phase is aqueous.

Ion exchange

Ion-exchange chromatography relies on the charge differences of the solutes for separation. The stationary-phase resins have a charged species associated with them that is displaced (or exchanged with) by the solute species. Figure 2.7 shows a schematic of the ion-exchange process for an anion-exchange resin. The negatively charged protein displaces the chlorine anions because of its greater negative charge at the pH of the mobile phase. By changing the mobile phase pH, the charge on the solutes can be manipulated to displace the ion, or so that the solute is displaced by the ion. Because of the sensitivity of a protein to pH, and a processing goal to minimize salt additions to a process, buffers that can adequately buffer at less than 0.05M are commonly used. Example buffers used for ion-exchange chromatography include: Tris, phosphate, acetic acid, and triethanolamine.

Gel filtration (size exclusion)

Gel filtration is a chromatography method that separates molecules by size on the

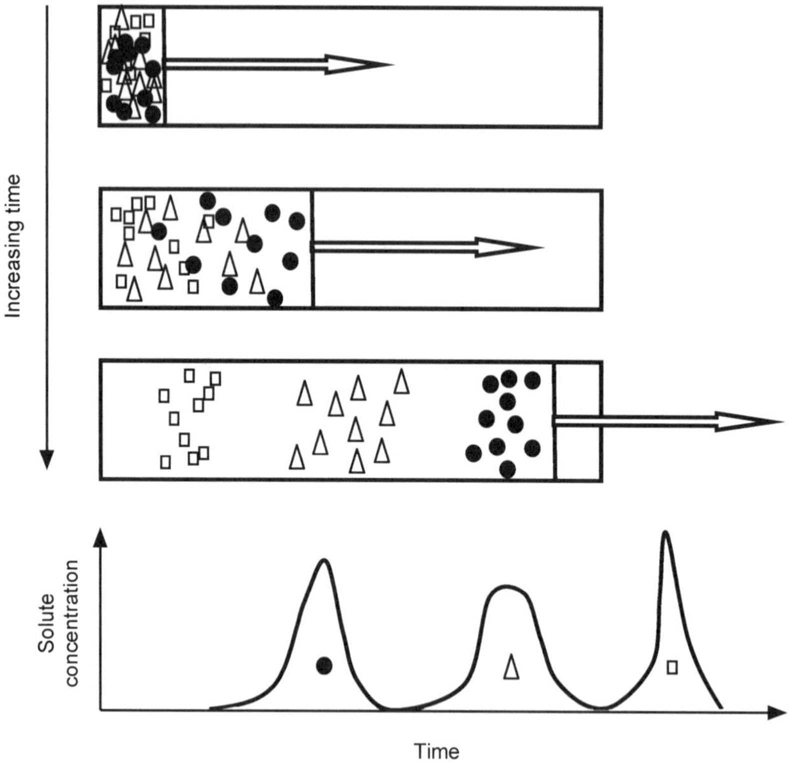

2.6 Schematic of elution chromatography. The three solutes exit the column at different times corresponding to the elution buffer concentration, where the equilibrium favors the fluid over the stationary phase.

same size-scale as ultrafiltration. Unlike ultrafiltration, the molecules are not retained by a filter media or membrane, but pass through the column packed with a stationary phase resin that is a soft spherical gel particle. The rate at which a molecule travels through the packed bed (stationary phase) is inversely dependent on size, i.e. large molecules exit first. The size exclusion resins are porous, such that small molecules have longer path lengths and, thus, take longer to go through the column. Figure 2.8 shows a diagram of how this system operates at the resin-scale for a small molecule and a protein. For therapeutic proteins, gel filtration is one of the few chromatography methods that does not require an additional step to remove the separating agent from the product stream.

Hydrophobic

Just as all proteins have charge, all proteins have hydrophilic and hydrophobic tendencies. It is because of these hydrophobic characteristics of a protein that the

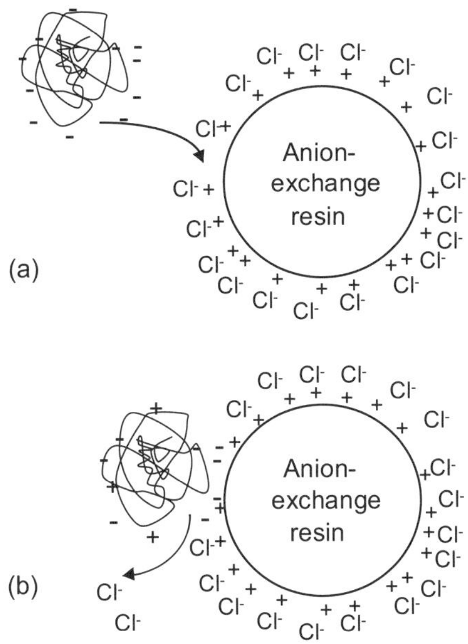

2.7 Schematic of the ion-exchange process for an anion-exchange resin: (a) negatively charged protein displaces the chlorine anions due to its greater negative charge at the pH of the mobile phase. (b) The elution buffer pH decreases, which causes the protein to become more positively charged so that it is displaced by the chloride ion.

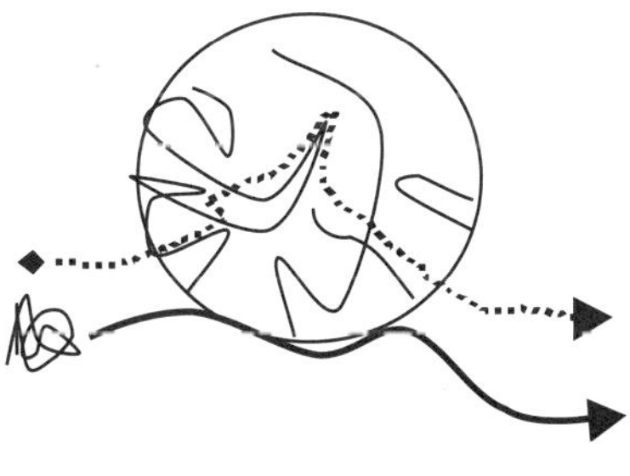

2.8 Schematic of a gel filtration resin with path length for two molecules of different sizes.

target protein can be separated from contaminant proteins. The proteins are adsorbed on to the stationary phase under very high salt concentration. The adsorbed proteins are desorbed as the salt concentration is decreased in order of decreasing hydrophobicity. Salt removal from the target solution is required after this step.

Affinity

Affinity chromatography relies on the specific interactions of a solute with a ligand. The ligand is attached to a support resin. The specific interaction between the solute and ligand is strong, similar to substrate–enzyme interactions. The affinity may be based on molecular size and shape. The forces controlling the interaction can include ionic, covalent, and/or hydrogen bonds. Most commonly, an antibody is used as the ligand, which recognizes the target protein with 'lock-and-key' precision. Affinity chromatography is the most expensive chromatographic method, since often a highly purified protein (the antibody) must also be manufactured before the target protein. Owing to the highly specific binding affinities, affinity chromatography can be used to give highly purified proteins in a single step, if the feed material is relatively free of cellular proteins, such as can be obtained from mammalian cell cultures grown in protein-free media. For biologically inspired textile protein, this method is too expensive to be considered practical or economically viable.

2.5 Finishing steps

Finishing steps are very dependent on the end-use of the protein and the protein characteristics of the final product. For example, some proteins denature under mild heat, so drying has limited use. Other proteins will not form crystal structures due to the amorphous nature of the protein, thus limiting crystallization. Finishing steps do not purify the target protein, however, but often significantly concentrate the protein.

2.5.1 Drying

Drying or precipitating concentrates a solution and transforms the solution into a solid by applying heat to evaporate the solvent, usually water. The solids formed normally have poorly defined morphology and small particles. This method is relatively inexpensive. Since many proteins are susceptible to heat, may have only limited application to protein solutions.

2.5.2 Freeze drying

Freeze drying or lyophilization is basically a drying operation, except the solvent, usually water, is removed by sublimination of the solvent from the solid to the vapor phase. Often a vacuum chamber is used to increase the speed of this process and allows for lower temperatures to be used, a gentler process for heat-susceptible proteins. The resulting solid has a relatively low density and re-dissolves readily back into the solvent. This method has been successfully used on enzymes and bacterial suspensions, where biological activity was recovered once the enzyme or bacterial were re-hydrated. Freeze drying may be appropriate for biologically inspired textile proteins as a means to concentrate the protein and then re-dissolve it in a solvent appropriate for the fiber formation steps.

2.5.3 Crystallization

Crystallization is a method for transforming a solution into a solid, where a supersaturated solution nucleates the solute by a chemical equilibrium controlled process. Uniform particles with well-defined morphology are formed, and these readily re-dissolve. Crystals tend to be brittle. Amorphous materials are more difficult to crystallize than highly ordered structures. Protein crystallization is more of an art than a science, but does allows researchers to determine the structure of a protein based on x-ray diffraction patterns. Due to the amorphous nature of many of the biologically inspired textile proteins, such as silk, crystallization may have only limited use.

2.6 Conclusions and sources of further information and advice

This chapter briefly covers the field of bioseparations as related to biologically inspired textile proteins. For economic considerations and slightly more in-depth coverage of the various unit operations, see *Bioseparations Science and Engineering* by Harrison *et al.* (2003). *Bioprocess Engineering* by Shuler and Kargi (2002) provides some background biochemistry and cell physiology, as well as detailed descriptions of the various separation and purification methods. An advanced-level, comprehensive coverage of bioseparations can be found in Ladisch's *Bioseparations Engineering* (Ladisch, 2001).

2.7 References

Harrison RG, Todd P, Rudge SR and Petrides DP (2003), *Bioseparations Science and Engineering,* New York, Oxford University Press.

Ladisch MR (2001), *Bioseparations Engineering,* New York, Wiley-Interscience.

Shuler ML and Kargi F (2002), *Bioprocess Engineering,* Upper Saddle River, NJ, Prentice Hall PTR.

3 Spinning of fibers from protein solutions

F. TEULÉ
University of Wyoming, USA

Abstract: The development of spinning technologies adapted for the processing of pure silk, collagen or elastin fibrous protein solutions made from natural and artificial fibrous protein sources is outlined. The natural silk-spinning mechanisms of spiders and silkworms are described to stress the physical and physiological requirements paramount to mimic the proper processing and spinning of pure fibrous protein solutions. Examples of the artificial production of collagen, elastin and silk fibers using wet-spinning or electrospinning technologies of pure protein solutions in different solvents as well as the properties of the materials produced are reviewed.

Key words: spinning technology, protein-based fiber, fibrous proteins, spider silk, silkworm.

3.1 Introduction

Fibrous materials are extremely prevalent in nature and their main components are usually polysaccharides or proteins. The industrial production of cellulose-based fibers is already a source of 'natural' polysaccharide-based fibers. Fibrous proteins, however, represent another interesting facet of 'natural' materials that deserves further exploration. The *in vivo* formation of some of these protein-based fibrous materials often involves 'spinning'. As a first step toward 'mimicking' natural protein-based fiber production, a thorough understanding of the native mechanisms of such fibrous protein processing is required. Therefore, the first part of this chapter is devoted to natural spinning mechanisms, where, after citing a few examples of 'spun' and 'non-spun' protein fibrous biomaterials, the *in vivo* silk-spinning mechanisms of orb-weaver spiders and silkworms are reviewed. The second part deals with the possible requirements to realize such artificial spinning of fibers from protein solutions outlining the main obstacles to overcome. Third, the pioneer work and current studies focused on the development of such spinning technologies are examined considering only the spinning from pure protein solutions. Finally, before concluding this chapter, the possible applications of synthetic protein-based fibers and possible future trends in this type of artificial spinning are addressed.

3.2 *In vivo* or natural spinning of protein-based fibers

In this chapter, the word 'spinning' refers to the specific process, performed at will, allowing the transformation of soluble fibrous proteins into an organized fiber or biofilament requiring the use of specialized apparatus. However, in nature, the production of such protein-based fibrous materials may not always require 'spinning'. Yet, biomaterials with exceptional properties are produced and, thus, their protein components are worth investigating as they too represent potential candidates for artificial spinning technologies.

3.3 Protein-based fibrous materials

3.3.1 'Non-spun' materials

Elastin and collagen are particularly versatile and may be the most prominent natural fiber proteins. Both elastin and collagen are found in vertebrates and collagen is also present in invertebrates. For instance, elastins are key components of elastic ligaments, skins or leathers, and blood vessels, whereas collagens are found in bones, cuticles, connective tissues, tendons, teeth, and ligaments. In vertebrates, both elastin and collagen proteins are components of the heavily crosslinked insoluble supramolecular structure known as the extracellular matrix where they assume specific roles (Indik *et al.*, 1987; Debelle and Alix, 1999; Wallace and Thompson, 1983; Ushiki, 2002). In the extracellular matrix, the collagen fibrillar system sustains the scaffold of cells and tissues while the microfibrillar elastin system homogeneously disperses stress to preserve the resilience adapted to local tissue requirement (Ushiki, 2002). Interestingly, these protein matrices are formed immediately following synthesis and secretion of the fiber proteins outside specialized cells displaying their ability to spontaneously self-assemble when reaching the extracellular compartment. This aggregation process occurs at several levels. First, nanofibrils form and self-organize into microfibrils leading to fibril formation which eventually yield a higher order structure. Several enzymes participate in this 'self-assembly process' and help stabilize the matrix by crosslinking the molecules and fibrils (Debelle and Alix, 1999; Wilson *et al.*, 1998; Edens *et al.*, 2001). In the invertebrate group of jawless fish, a different type of self-aggregating fibrous protein can be found in the extracellular matrix. For example, in sea lampreys, some of the cartilaginous structures are composed of collagenous matrices, while others are made of non-collagenous and non-elastin like small proteins called 'lamprins' (Robson *et al.*, 1993). In some insects such as locusts and dragonflies, another kind of elastic fibrous protein called 'resilin' is a component of specialized regions of the cuticle as well as joints and tendons (Weis-Fogh, 1960; Weis-Fogh, 1961). Resilin contains structural motifs resembling those found in elastins and is also crosslinked through covalent bonds (Andersen, 1964). Its role is to provide long range elasticity to some tissues and, during insect flight, resilin allows energy storage and acts as a damper of vibrations.

Vertebrate muscles also contain numerous fibrous proteins. The giant titin protein, found in sarcomeric filament systems in striated muscles of higher vertebrates, is involved in muscle assembly and provides tissue elasticity (Tskhovrebova and Trinick, 2003).

Finally, keratin proteins also yield heavily crosslinked fibrous materials and are another interesting example of biomaterials. Here again, no obvious spinning mechanism is used but rather a controlled self-aggregation process that, in one case, yields true fibers such as hair or wools while in others forms more compact fibrous structures like nails, horns or quills (Chapman, 1969; Feughelman, 2000).

3.3.2 'Spun' materials

Very few organisms seem to have the ability to deliberately spin fibrous proteins into fibers using specialized spinning apparatus but many arthropods do possess such skills. This spinning process allows for the accumulation and storage of the self-aggregating fibrous proteins in a soluble form until their assembly is initiated to form the final water-insoluble fiber.

Marine mussels for instance attach themselves to rocks and to each other in order to survive in tidal waters using protein-based fiber anchors. These byssal threads or 'beards' are external collagen-based tendon-like structures. These tough holdfasts are 'spun' and glued to a hard surface, underwater by the foot of the mussel (Pujol, 1970; Waite, 1992). Although the nature and the repartition of the proteins making up the byssus are known (Qin *et al.*, 1995; Qin *et al.*, 1997; Coyne *et al.*, 1997; Qin and Waite, 1998; Waite *et al.*, 1998; Coyne and Waite, 2000), the actual spinning process used to generate these byssal threads is not yet fully understood but seems to resemble injection molding (Waite, 1992).

Insects and arachnids also fabricate protein-based fibers at will using spinning apparatus. Prominent examples are silkworms (insects, Lepidoptera) and spiders (arachnids, Araneidae) which are able to spin water-insoluble dry silk filaments from a liquid silk protein solution (or 'silk dope'). These fibers have diameters from 2 to 5 μm for spiders' silks and 10 to 20 μm for silkworms' silks, and are used by these organisms as external gear in their daily life. The self-assembly of silk proteins seems to require some still uncharacterized processing and the use of sophisticated and tightly regulated spinning apparatus to create fibers. Silks are not simply extruded or squeezed outside the organism's body but rather are drawn out by the spider's legs (Foelix, 1996) or by the silkworm's figure-eight-shaped head movements (Akai, 1983). Silks are of great interest because of the wide range of their mechanical properties due to the widespread variety of their uses (Craig, 1997).

The sequences of the proteins constituting the silks of silkworms (Kusuda *et al.*, 1986; Kikuchi *et al.*, 1992; Mita *et al.*, 1994; Zhou *et al.*, 2000; Datta *et al.*, 2001) and many spiders (Xu and Lewis, 1990; Guerette *et al*., 1996; Hayashi and Lewis, 1998; Colgin and Lewis, 1998; Hayashi and Lewis, 2000; Gatesy *et al*., 2001; Tian and Lewis, 2004) are available. The data suggest that these proteins are highly repetitive, usually contain high amounts of glycine, serine and alanine, and share several structural motifs (hydrophilic amorphous motifs and hydrophobic crystalline forming motifs) that are believed to play specific roles in mechanical properties (Gosline *et al*., 1999; Hayashi *et al*., 1999). For example, dragline and flagelliform silks, two components of orb-weaver spider webs (frame and radius, and catching spiral, respectively), are made of large silk proteins with distinct primary structures (MaSp 1 and MaSp 2, and Flag, respectively) that share similar structural motifs but in different combinations and amounts (Xu and Lewis, 1990; Hinman and Lewis, 1992; Hayashi and Lewis, 1998; Hayashi *et al*., 1999; Hayashi and Lewis, 2000). As a result, dragline and flagelliform silks, both extremely tough fibers (160 and 150 $MJ.m^{-3}$, respectively), exhibit different tensile strengths (σ_{max} = 4 and 0.5 GPa, respectively) and elasticities (ε_{max} = 35 and 200%, respectively) specifically adapted to their roles (Gosline *et al*., 1984; Gosline *et al*., 1999; Denny, 1976; Stauffer *et al*., 1994). In comparison, *Bombyx mori* silkworm silk combines a tensile strength of 0.61–0.74 GPa and an elasticity of 15–24% (Denny, 1980; Perez-Rigueiro *et al*., 2000).

Interestingly, while elastin-, collagen-, and keratin-based fibrous materials are stabilized through a combination of covalent and non-covalent interactions, some silks are only stabilized through non-covalent interactions. Moreover, it is worth mentioning that silks proteins in general are significantly larger proteins than other self-assembling fibrous proteins. The sizes of both dragline protein monomers (MaSp 1 and MaSp 2), estimated by gel filtration, are 300–350 kDa (Sponner *et al*., 2005). The heavy-chain fibroin from silkworm silk (*Bm-Fhc*) is also roughly 350 kDa. However, in silkworm silk, the heavy-chain fibroin is crosslinked through disulfide bonds to the 25 kDa light-chain fibroin (*Bm-FLc*) protein and the *Fhc/Flc* fibroin complex is non-covalently associated to the 27 kDa P25 glycoprotein (Tanaka *et al*., 1999a; Tanaka *et al*., 1999b). In comparison, the sizes of collagen protein monomers are usually 30–100 kDa (Kaddler *et al*., 1989; Cox, 1992; Kramer, 1994; Ray *et al*., 1996; VanderEycken *et al*., 1994; Wang *et al*., 1998).

3.4 Silk production in spiders and insects: a natural spinning process

Silk production and spinning processes in both silkworms and spiders are initiated in specialized internal organs or silk glands that are linked to external spinneret structures through essentially what are short or long tapering ducts. The spinnerets are covered by one or more openings functioning as spigots through which the solidified fiber exits. Thus, silk production is achieved under physiological conditions in a regulated aqueous environment, at ambient temperature as well as

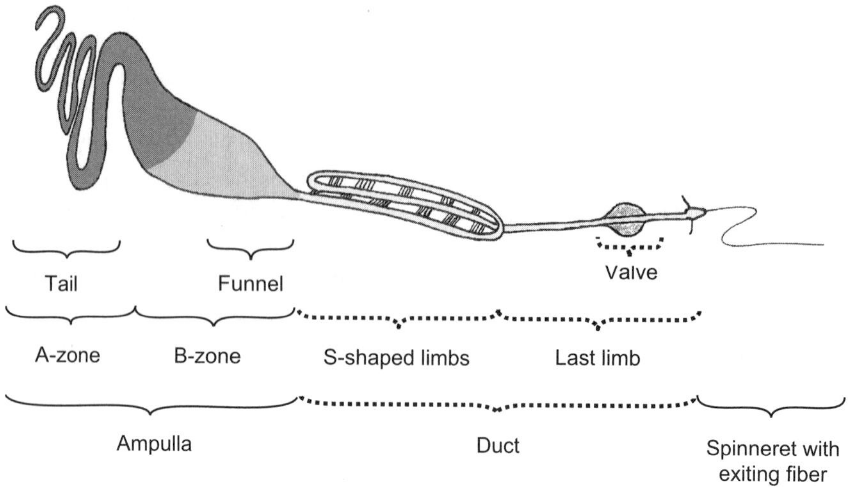

3.1 Major ampullate gland spinning system in orb-weaver spiders (*Nephila edulis*). After Vollrath *et al.* (1998) and Vollrath and Knight (1999 and 2001).

low hydrostatic pressure and with relatively slow extrusion/drawing rates. While displaying evident anatomical differences, the silk-producing systems of spiders and silkworms seem to share similarities in their design and mode of operation.

3.4.1 Orb-weaver spiders

Orb-weaver spiders: spinning apparatus

The silk glands are located in the spider's abdomen and the spinnerets are found externally on the ventral side of its lower abdomen (Lucas, 1964; Foelix, 1996). The spinning systems of several spiders have been investigated (Kovoor, 1977; Kovoor, 1986; Kovoor, 1990; Vollrath and Knight, 1999). The most recent reports thoroughly describe the major ampullate spinning system of the *Nephila edulis* (Araneae) orb-weaver responsible for the production of the major ampullate silk used as dragline and web bearing frame (Vollrath *et al.*, 1998; Vollrath and Knight 1999).

In this system, the ampulla is divided in two zones (A and B) lined with secretory epithelium (Fig. 3.1). The A-zone (proximal) is composed of a long tail and the first part of the sac while the B-zone (distal) comprises the widest part of the ampulla up to the funnel (constriction). Previous histochemical data suggested that the nature of the secretions in the two zones differed (Kovoor, 1986) and it was initially thought that most protein secretions in the A-zone constituted the core of the silk fiber. In this zone, tiny spherical viscous droplets are visible which gradually increase in size, most likely coalescing with one another. As they move forward in the ampulla towards the duct region, the growing droplets seem to be stretched by the elongational flow (Knight and Vollrath, 1999). In the B-zone, the secreted protein mixture seemed to coat the fiber (Vollrath and Knight, 1999). Recent immunological data determined that MaSp 1 and MaSp 2, both dragline silk proteins, are produced and secreted throughout the A- and B-zones of the gland (Sponner *et al.*, 2005). Thus, both proteins accumulate in the gland lumen and constitute the core of the fiber while a third protein coats the nascent fiber. In fact, a glycoprotein, detected in both gland (Kovoor, 1977) and thread (Weiskopf *et al.*, 1996), may constitute the coating (Vollrath and Knight, 2001) and plasticize the thread by retaining moisture (Vollrath and Tillinghast, 1991). An increasing gradient of peroxidase from the proximal B-zone to the region just before the funnel was also detected though the role of this enzyme remains obscure since silks do not contain any di- and tri-tyrosine linkages. The silk material exits the gland through a funnel and flows into a long narrow S-shaped (first limbs of the duct) tapering duct ending with a valve located a few millimeters before the spinneret (Fig. 3.1). Surfactants like lipids are added in the last stages (last limb of the duct before the valve) of the spinning process. The S-shaped duct is characteristic of major and minor ampullate glands spinning systems while such a valve is only found in the duct of major ampullate glands (Vollrath *et al.*, 1998). Owing to the nature of its cuticle, the duct functions like a dialysis tubing, thus silk dope's pH as well as salt and surfactants concentrations allowing water removal, are regulated at different positions along the spinning duct system during native spinning (Knight and Vollrath, 2001; Vollrath and Knight, 2001). Indeed, staining techniques followed by scanning electron microscopy–energy dispersive spectroscopy (SEM-EDAX) demonstrated a decrease in pH from the ampulla tail region to the end of the duct just right before the spigot (7.2 to 6.3). In addition, the presence of a proton pump in the third limb of the duct showed evidence of a final acidic treatment of the material (Vollrath *et al.*, 1998). Finally, changes in ion concentrations along the duct are such that there is a decrease in sodium ions (Na^+) concurrent with an increase in the more chaotropic potassium ions (K^+).

Orb-weaver spiders: spinning mechanism

Data from magic angle spinning nuclear magnetic resonance (MAS NMR) indicated that the silks proteins within the sac of the major ampullate gland of *Nephila edulis* [concentration of 30–40% (w/v)] constitute an isotropic phase (Hronska *et al.*, 2004). However, at concentrations greater than those existing in the gland, the silk dope behaving as a lyotropic liquid crystal enters a nematic phase (Kerkam *et al.*, 1991; Willcox *et al.*, 1996; Viney, 1997; Knight and Vollrath 1999; Vollrath and Knight, 2001). During spinning, this lyotropic liquid crystal goes through a gel-like state that is then converted into a dry solid fiber. This liquid crystalline phase plays a rheological role in the spinning process and therefore impacts the mechanical properties of the spun fiber (Viney, 1994; Viney, 1997; Vollrath and Knight, 2001).

This state seems to arise partially because silk proteins are amphiphilic molecules alternating hydrophobic and hydrophilic motifs (Vollrath and Knight 2001; Kerkam *et al.*, 1991). Thus, in the aqueous environment of the gland and duct, and in response to appropriate changes in solvent conditions (acidification, and increase in K^+ concurrent with decrease in Na^+), the silk proteins which were in a soluble storage conformation, supposedly a mix of random coils and helices, start to unfold exposing hydrophobic residues. These 'surfacing' hydrophobic residues thermodynamically trigger the coalescence of the silk proteins through a structural transition to crystalline β-sheets resulting in exclusion of water, and thus phase separation (Knight and Vollrath, 2001; Chen *et al.*, 2002; Braun and Viney, 2003; Dicko *et al.*, 2004a; Dicko *et al.*, 2004b). Coincident with these chemical changes in the dope, stress-induced forces orienting the protein chains are also critical in bringing about this transition leading to phase separation (Vollrath *et al.*, 1998; Knight *et al.*, 2000; Knight and Vollrath, 2001). Studies relying on Congo red staining determined that the origin of these stress-induced forces was mostly due to elongational flow in the gland and duct rather than shearing from wall friction (Knight *et al.*, 2000). Such elongational flow is a direct consequence of the geometry of the gland and duct which combines a funnel giving into a slow long narrowing duct that functions as a hyperbolic die with rapid internal draw-down taper occurring in the third limb of the duct before the valve. Phase separation occurs at the beginning of the draw-down taper where β-sheet formation seems to be initiated, thus it is at this position that a solid fiber thread surrounded by an aqueous phase becomes visible (Knight *et al.*, 2000). Moreover, lipids added in the final part of the duct where the draw-down occurs, may also favor protein unfolding in solution acting as plasticizers (Vollrath and Knight, 2001). Evaporation may lead to additional water loss when the spider finally draws the fiber into the air (Vollrath and Knight, 2001; Willcox *et al.*, 1996).

3.4.2 Silkworms

Silkworms: spinning apparatus

The domesticated *B. mori* silkworms (Lepidoptera, Bombycidae) possess one pair of silk glands (modified salivary glands) arranged symmetrically on the ventrolateral sides of the mid intestine of the insect (Akai, 1983). These glands are each connected to a short tapering duct. Both ducts eventually unite (Y-junction) into a single large spigot constituting the spinneret mounted on the base of the labium just posterior to the mouth of the insect (Tnau, 2003). A single silk gland contains distinct sections assuming different functions: posterior, median (three regions), and anterior. The posterior region of the gland secretes the silk fibroin dope (heavy- and light-chain fibroin proteins) while the three distinct middle parts secrete the protein coating the fiber (sericin) and also serve as a reservoir for the storage of the silk fibroin proteins (Kikuchi *et al.*, 1992; Mita *et al.*, 1994; Grzelak, 1995; Zhou *et al.*, 2000; Datta *et al.*, 2001). The silk dope moves from the posterior region towards the anterior region of the gland before flowing into the tapering duct. Since both ducts coming from each gland are joined before reaching the single press placed right before the spigot, the silk fiber drawn out of the silkworm's mouth is formed of two individual coated filaments (Magoshi *et al.*, 1985a, 1985b; Sehnal and Akai, 1990). The press may act as a restriction die able to regulate diameter (Asakura *et al.*, 2007). The pH decreases from the proximal part of the gland towards the end of the duct (from 6.9 to 4.8).

Silkworms: spinning mechanism

Although extensively investigated (Akai, 1983; Magoshi *et al.*, 1985a, 1985b; Asakura *et al.*, 2007), the spinning process in silkworms is not completely understood. Like spider silk processing mechanisms, studies of *B. mori* native silk dope also suggest the existence of a lyotropic liquid crystalline phase (Magoshi *et al.*, 1985a; Nakamae *et al.*, 1989; Kerkam *et al.*, 1991), as well as shear forces, elongational flow, and existence of a draw-down taper that are able to promote phase separation of the silk dope (Asakura *et al.*, 2007). In parallel with these stress-induced forces, additional factors such as changes in protein composition and concentration, pH regulation and changes in ion concentrations (K^+, Na^+ and apparently mainly an increase in Ca^{2+}) (Zhou *et al.*, 2000; Zhou *et al.*, 2003; Wong Po Foo *et al.*, 2006a; Asakura *et al.*, 2007) that affect protein conformational transitions and water content also play a decisive role in phase separation leading to fiber formation (Asakura *et al.*, 2007). Here again, the lyotropic liquid crystalline silk dope enters a nematic phase in the region of the draw-down taper right before the silk press (Asakura *et al.*, 2007). Indeed, the birefringence increases dramatically in the internal draw-down taper suggesting an increase in molecular orientation (Asakura *et al.*, 2007). In conclusion, a change in silk protein conformation from soluble silk (silk I) to insoluble silk (β-sheet or silk II) seems to be initiated in the spinning duct before the press at the drawn-down taper site (Asakura *et al.*, 2007).

3.5 Elements to consider for the *in vitro* or 'artificial' spinning of protein-based fibers

After reviewing known natural spinning mechanisms involved in silk production, several key factors seem to be critical in promoting fibrous proteins self-assembly and thus should be considered when developing artificial spinning technologies. In this section, the details of the critical factors required in natural spinning are explored and their possible management in the development of artificial spinning technologies is discussed.

3.6 Factors involved in native self-assembly processes

Important factors involved in fibrous proteins self-assembly seem to be protein primary structure, solvent conditions, and, in the case of spun fibers, the design of the spinning apparatus. As mentioned earlier, these soluble fibrous protein monomers possess a self-organization capability, which, under appropriate aqueous solvent conditions and processing, allows for the formation of the final, sometimes crosslinked, insoluble supramolecular structure. Additionally, all of these factors have a noticeable impact on the mechanical and physical properties of the materials formed.

3.6.1 Protein primary structure and choice of solvent

The fact that the primary sequences of these diverse natural fiber proteins rely on a set of specific structural motifs stresses the importance of the role of polypeptide primary sequence. In fact, sequence similarities between many natural fibrous proteins helped to assign structure–function relationships. Hypotheses regarding the secondary structures of certain amino acid motifs comprising some of these fibrous proteins as well as their possible role in the mechanical properties of the final supramolecular matrices or fibers were formulated (Urry *et al.*, 1984; Hayashi *et al.*, 1999; Tatham and Shewry, 2002). For example, proline-containing pentapeptides found in elastins and some orb-weaver spider silks or poly(A)/poly(GA) [poly(alanine)/poly(glycine–alanine)] motifs identified in silks adopt distinct secondary structures (β-turns/β-spirals or β-sheets, respectively) inherent to their primary structures.

Such types of structural motifs in fiber proteins are thought to grant elasticity or strength, respectively, to the final fiber or matrix produced (Gosline 1978; Urry *et al.*, 1986; Gosline, 1987; Urry, 1988; Hayashi *et al.*, 1999). Furthermore, as described earlier in the native silk spinning process, the secondary structures adopted by the motifs present in these fibrous proteins are flexible thus may change under different chemical and physical conditions stressing the importance of the solvent and its intricate relationship with the proteins. Indeed, structural transitions resulting in protein conformational changes (refolding) in response to an evolving surrounding environment subjected to various extrinsic factors can be such that a soluble protein may suddenly become insoluble under specific conditions. In the case of spider dragline and silkworm silks, this type of change in protein conformation resulting from the structural transition of poly(A) or poly(GA) motifs from random coil and/or helical to crystalline β-sheets seems necessary to initiate the coalescence of the silk proteins. However, this functional role in self-assembly is not just restricted to alanine-containing motifs.

Recent studies on several elastin structural motifs suggested that specific glycine-rich sequences devoid of proline may behave in an amyloidogenic manner thus promoting aggregation through structural transition involving β-sheet formation (Rausher *et al.*, 2006). In fact, such types of glycine-rich sequences are found in many self-aggregating proteins which may or may not contain any poly(A) and/or poly(GA) sequences. More specifically, some of these glycine-rich sequences found in elastins are also present as $(GGX)_n$ tripeptides in several silks (Colgin and Lewis, 1998; Xu and Lewis, 1990; Hayashi and Lewis, 1998), or as $(GGLGY)_n$ pentapeptides in lamprin proteins (Robson *et al.*, 1993; Bocchochio *et al.*, 2001). As a consequence, the artificial spinning of fibrous proteins appears to be more challenging than the spinning of a simple uniform chemical polymer because of the complexity and heterogeneity of the polypeptide primary structure and also its behavior in different solvent conditions, creating complex yet critical intramolecular and intermolecular interactions between amino acids of the chains.

3.6.2 Spinning technologies

As mentioned in the case of natural spinning, besides protein sequence and chemical modification of the dope, the design of the spinning device is also fundamental to promote fiber formation. Indeed, additional stress-induced forces caused by elongational flow and shearing help in phase transition and separation in native silk processing (Vollrath *et al.*, 1998; Knight *et al.*, 2000; Knight and Vollrath, 2001). Thus, applying conventional spinning methods to spin synthetic fibers from aqueous solutions is more challenging. Besides the fact that these spinning techniques were developed for industrial large-scale production of synthetic chemical polymer fibers and thus are not particularly suited for small-scale recombinant production, these systems are somewhat 'closed', consequently, they do not offer the option of progressive chemical modification of the spinning dope to help in structural transitions.

It is also important to keep in mind that in *in vivo* spinning processes, elongational flow and spinning occur simultaneously and are important in promoting molecular orientation. However, in an artificial spinning system, depending on the technique used, elongational flow may or may not occur during spinning consequently extra post-spinning treatments are required to improve the mechanical properties of the 'as-spun' fibers.

3.7 'Mimicking nature'

To generate protein fibers, the starting protein materials used are 'native' if extracted from natural materials (animal tissues, silks) or 'synthetic' if produced as recombinant protein analogs. With recombinant protein production, the fibrous protein gene may be cloned as cDNA or as 'true synthetic' gene (mimetic). Sometimes, the molecular weights of native and recombinant proteins may be different. In the case of silks, the synthetic analogs will most likely be smaller than the native versions and the repeat structure of a true synthetic will be stricter than that found in natural silks. Synthetic fibers made from smaller analogs may have limited overlaps of molecular chains, thus, fewer possible stabilization interactions and perhaps smaller crystals than those found in native silks. As a result, the mechanical properties will be affected. It is critical to investigate the nature of these interactions in order to design analogs capable of promoting optimized molecular interactions. Usually, native or synthetic materials are available as lyophilized proteins and the first step in artificial spinning comprises solubilizing these in appropriate solvents that will prevent their premature aggregation. Once again, the choice of the solvent may vary depending on the nature and condition of the material.

In developing artificial spinning technologies, it is fundamental to understand how to control the protein unfolding and refolding processes necessary to create the proper protein structural transitions leading to self-assembly. Sequences known to promote 'self-assembly', like crystalline-forming motifs, should be considered when designing synthetic fibrous protein genes. However, the lyotropic nature of these types of molecules renders the handling of the protein solution more difficult due to the tendency of the peptides to aggregate at a specific concentration threshold. Studies showed that addition of chemically or enzymically controlled trigger sequences allowed control over self-assembly of silk analogs (Winkler *et al.*, 1999; Winkler *et al.*, 2000; Wong Po Foo *et al.*, 2006b). The recombinant silk proteins contained polyalanine motifs flanked by methionine (M) redox or phosphate (RGYS*L) trigger sequences. Oxidation of the methionines or the phosphorylation of the RGYS*L sequences increased the solubility of the silk analogs by sterically disrupting β-sheets. Self-assembly was initiated when reversing the state of the triggers (reduction or dephosphorylation). Many studies on silks demonstrated that other chemical factors involved in natural silk spinning could be used *in vitro* to promote self-assembly. Indeed, acidification (Dicko *et al.*, 2004c) and increase in concentrations of different cations (Dicko *et al.*, 2004c; Chen *et al.*, 2002) affected β-sheet transitions in silk aqueous dopes. In addition, temperatures (up to 85 °C) also promote several critical structural transitions in native (Dicko *et al.*, 2004a) and synthetic silk proteins (Teulé *et al.*, 2007) required for protein self-assembly in aqueous solutions. For spinning, temperature-induced structural changes, when applicable, would be easier and cheaper to implement than chemical or enzymic processes.

The main available spinning technologies used to spin fibrous protein materials initially relied on 'wet-spinning' (or 'dry-jet' wet-spinning) and more recently expanded to 'electrospinning'. Wet-spinning usually produces single fibers with rather large diameters (tens to hundreds of micrometers) whereas electrospinning, which is cheaper, generates nanometer-scale fibers as non-woven meshes. Thus, depending on the future application of the fibers, one technique may be chosen over the other. It is important to outline that elongational flow fields occur during electrospinning, while no elongational flow forces are present during wet-spinning. Therefore, the nature and extent of the post-spinning treatments necessary to improve mechanical properties may not be the same in both cases. These post-spinning treatments are essential to improve the overall assembly and internal structure of the 'as-spun' fibers usually by increasing crystallization (dehydration) and crystal orientation (drawing) thus maximizing molecular interactions. Once again, the nature and condition of the material along with the mode of spinning will dictate the types of appropriate post-spinning processes. Other spinning factors such as extrusion rates, choices of dope solvent and coagulation solvents are also important.

Therefore, from one study to another, spinning 'dope' preparations, spinning techniques and conditions, as well as post-spinning processes may differ to suit a particular fibrous protein. Consequently, the mechanical performances of the artificially spun and processed fibers may be highly variable.

3.8 Examples of protein-based fibers produced through artificial spinning technologies

Today, while films and hydrogels are generated from native or recombinant collagens, elastins and silks, spinning fibers from these fibrous proteins is still an area of active research. This section will therefore only focus on the studies reporting protein fiber production from pure protein materials.

3.9 Wet-spinning of fibrous proteins

3.9.1 What is wet-spinning?

Wet-spinning is a technology based on extrusion processes and is widely used in industry to produce synthetic and regenerated cellulose fibers. The manufacturing of synthetic fibers based on long-chain synthetic polymers such as acrylic, aramid or spandex, or natural polysaccharide polymers such as rayon (http://www.fibersource.com/f-tutor/techpag.htm) rely on wet-spinning techniques. In this mode of fiber production, the polymer solubilized in a solvent forms a viscous liquid substance which is pushed through the tiny holes of a spinneret in a process referred to as extrusion. The spinneret is immersed into a chemical bath to coagulate the emerging polymer filaments and form solid fibers (Fig. 3.2).

3.9.2 Wet-spinning applied to the production of protein-based fibers

Wet-spinning of collagen fibers

While natural collagen extracts have been used as biomaterials for a long time, only a few reports of successful trials regarding the wet-spinning of collagen fibers may be found in the literature. These studies rely exclusively on the use of

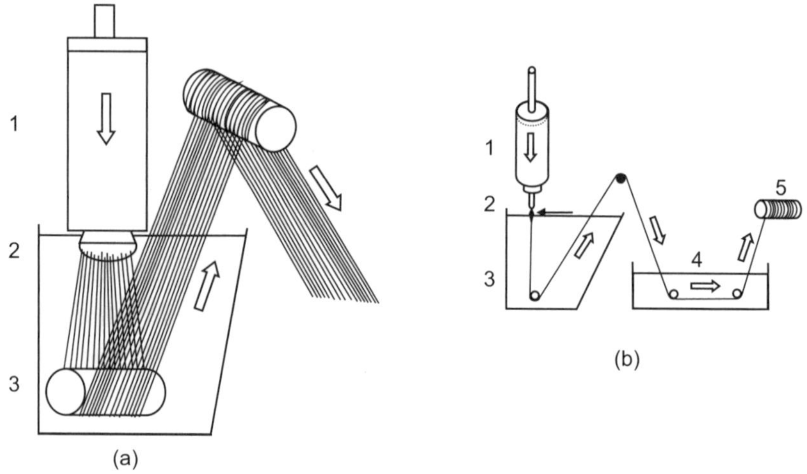

3.2 Wet-spinning techniques. (a) Example of industrial-scale wet-spinning set-up; (1) extruder containing the spinning dope; (2) shower-head type spinneret with multiple openings; (3) chemical coagulation bath. The solidified fibers are taken up by a bobbin. (b) Example of small-scale wet-spinning set-up for artificial production of protein fibers; (1) extruder containing the fibrous protein dope (i.e. syringe type extruder with mechanically or manually controlled plunger); (2) single-spigot spinneret (i.e. hypodermic needle or small peek tubing); (3) coagulation bath with single pully taking up the solid fiber; (4) example of post-spinning drawing treatment performed in solvent. Two godets rotate at different speeds to stretch the fiber (filled arrow); (5) the processed fiber is collected on a winder.
In (a) and (b), the spinning dope is pumped through the spinneret and solidifies upon contact with the coagulation bath (empty arrows). The solid fibers (thin lines) are directed towards the rest of the spinning line [not represented in (a)], where post-spinning processes may be applied (b). The technique used may be 'dry-jet' wet-spinning (b) when the spinneret is not immersed in the coagulation bath (2, black arrow showing air gap). For 'gel spinning', the dope is modified (1) so that its state is no longer a solution but not yet a solid (i.e. gel-like).

extracted natural animal collagen materials or gelatins (hydrolyzed collagen), which can be effectively processed into powders. In a patented aqueous wet-spinning process, collagen aqueous solutions of very low pH (pH <3) were extruded into a coagulation bath containing an inorganic salt (NaCl, $MgCl_2$, $Na_2(SO)_4$, $(NH_4)_2SO_4$, or $Al_2(SO_4)_3 16H_2O$) (Furukawa *et al.*, 1994). The dry fiber was consecutively crosslinked by treatment with formaldehyde or glutaraldehyde (Furukawa *et al.*, 1994). More recently, another patented procedure described the extrusion of a 0.05 wt% collagen–acetic acid solution into a heated (35 °C) alginic acid/boric acid bath (pH 8–10) (Fofonoff and Bell, 1999). In this particular case, the neutralization of the acidic dope by the alkaline bath initiated the self-assembly of the collagen fibrils into a fiber. The fibers which were successively dehydrated by treatments with acetone and ethanol had diameters in excess of 100 μm. Other studies showed that 20 wt% gelatin aqueous solutions could also produce fibers which were dried by heat-treatment (at 150 °C for 3 h) (Nagura *et al.*, 2002). The dry gelatin fibers achieved tensile strengths and initial modulus (σ_{max} = 130 MPa, $E_{initial}$ = 7 GPa) (Nagura *et al.*, 2002) comparable to that of some mammalian collagen tendons (σ_{max} = 120 MPa, $E_{initial}$ = 1.2 GPa) (Pollock and Shadwick, 1994).

More recently, dimethyl sulfoxide (DMSO) was investigated as an alternative solvent for gelatin (Fukae *et al.*, 2005). In this study, a 10 wt% gelatin–DMSO dope maintained at 60 °C was extruded into a methanol bath held at –20 °C in a gel-spinning process. The fibers were drawn immediately at different lengths and consecutively immersed in methanol for one week to remove the solvent before mechanical testing. Unprocessed fibers had low tensile strength and stiffness (σ_{max} = 28 MPa, $\sigma_{initial}$ = 0.7 GPa). While drawn fibers (having a draw ratio, or DR, of 4) had improved mechanical properties (σ_{max} = 81 MPa, $E_{initial}$ = 1.9 GPa), additional crosslinking with gluteraldehyde had little effect on their performance (σ_{max} = 112 MPa, $E_{initial}$ = 1.9 GPa). However, extensive post-spin drawing of the gelatin fibers (DR = 8) drastically improved both their tensile strength and stiffness (σ_{max} = 180 MPa, $E_{initial}$ = 3.4 GPa). X-ray diffraction data showed that drawing induced orientation of pseudo-crystallites present in the fiber (Fukae *et al.*, 2005).

Wet-spinning of silk fibers

The availability of degummed silkworm silk materials (silk fibroins) facilitated the regeneration of silks using wet-spinning. In 1993, the first process using wet-spinning to regenerate silk fibers from *B. mori* silks was patented (Lock, 1993). The method relies on the use of silk fibroins resolubilized in 1,1,1,3,3,3-hexafluoro-2-propanol (HFIP) as a spinning dope, which is extruded through a small needle into a methanol bath. The method was later used to regenerate silkworm silk fibers from 15 wt% silk fibroin-HFIP dopes to study the structural changes occurring during spinning and consecutive post-spinning drawing (Trabbic and Yager, 1998). The fibers were drawn into the air while still wet with methanol (DR = 1–3.5) and dried at constrained lengths (Trabbic and Yager, 1998). Circular dichroism and Raman spectroscopy data indicated that the fibroins which predominantly adopt distorted α-helix conformation in HFIP, mainly transitioned to β-sheet structures in the 'as-spun' fiber due to the methanol treatment. X-ray data showed that while the 'as-spun' fibers were more than 50% crystalline with little crystallite orientation, post-spinning drawing (DR ≥ 2.5) was critical in promoting crystallite orientation close to that observed for the native silk (Trabbic and Yager, 1998). Almost simultaneously, a miniature spinneret allowing the wet-spinning of meters of fibers from solutions containing as low as 10 mg of proteins was constructed using silicon technology (opening = 80–160 μm) (Liivak *et al.,* 1998). This microspinneret was used to spin native silkworm silk and later *Nephila clavipes* dragline spider silk. For silkworm silk, a 2.5 wt% silk-HFIP dope was extruded into methanol (Liivak *et al.,* 1998) following previous methods (Lock, 1993; Fahnestock, 1994). The fibers, drawn while wet, were successively soaked in methanol overnight before drying and annealing (at 40 °C for 1 h). The mechanical properties of the regenerated fibers were close to those of the native silk (Liivak *et al.,* 1998). NMR studies indicated that a decrease in aperture size with an increased draw ratio caused an increase in the β-sheet fraction constituted by the alanine domains. Additionally, maximum stresses were higher when both β-sheet fraction and crystal orientation were increased (Liivak *et al.,* 1998). Conformational transition to β-sheet structures and successive molecular orientation of the formed crystallites occurred as a result of these spinning/post-spinning processes (Jelinski *et al.*, 1999). For spider silk though, the 2.5 wt% silk-HFIP dope had to be extruded into acetone to regenerate fibers (Seidel *et al.*, 1998; Seidel *et al.*, 2000). Under SEM, the brittle regenerated fiber appeared spongier than the native spider silk. NMR data of the dry 'as-spun' and water-treated fiber indicated that subsequent water-treatment increased the fraction of alanine residues adopting β-sheet structures (Seidel *et al.*, 1998).

Later, the same team used a two-step post-spinning procedure during which the fibers were first drawn into air while wet with acetone, dried at constrained length, and then soaked in water before being stretched again while still wet (Seidel *et al.*, 2000). Mechanically, these double-drawn fibers outperformed the single-drawn and as-spun fibers (σ_{max} = 320 MPa, $E_{initial}$ = 8.0 GPa) (Seidel *et al.*, 2000). These values, closer to those of native dragline silk, were far superior to the single air-drawn fibers made with synthetic spider silk analogs (Fahnestock, 1994). These experiments established the importance of the role of water in the post-spinning processes. Silkworm silks were also successfully spun using 10 wt% silk fibroin–hexafluoroacetone hydrate (HFA) dopes and methanol as a coagulant (Yao *et al.*, 2002). The drawn (DR = 3) and steam-annealed (at 125 °C for 30 min) fibers displayed an initial modulus and a tensile strength comparable with and half that of native silk, respectively (Yao *et al.*, 2002).

Although these initial studies using organic solvents are encouraging, their high price and toxicity are problematic for any eventual large-scale fiber production, therefore alternative solvents have been investigated. Recently, silk fibroins were extruded from formic acid (FA) solutions using 13–19% (wt/v) dopes (Um *et al.*, 2004; Ha *et al.*, 2005), a solvent used to artificially generate a fiber from a recombinant MaSp 2 protein (Lewis *et al.*, 1996). The regenerated silkworm silks (drawn while wet) had interesting properties ($\sigma_{max} \approx$ 1 GPa, ε_{max} = 30%), yet again demonstrating the impact of fiber processing on mechanical performances (Um *et al.*, 2004; Ha *et al.*, 2005).

N-Methylmorpholine *N*-oxide (NMMO) monohydrate, a recyclable non-toxic solvent, was used to regenerate silkworm silks with improved mechanical properties (Marsano *et al.*, 2005). The 13 wt% silk fibroin–NMMO solution was extruded into ethanol. The data showed that an increase in draw ratios (DR_1/fiber take up = 1–15 and DR_2 = 1–2.7) resulted in a decrease in fiber diameter (130 to 19 μm), concomitant with increases in birefringence, initial modulus ($E_{initial}$ = 2.6–7.2 GPa), tensile strength (σ_{max} = 43–120 MPa) and elasticity (ε_{max} = 2–35%) (Marsano *et al.*, 2005). The regenerated fibers still exhibited initial modulus and tensile strength lower than those of native silkworm silk ($E_{initial}$ = 15–17 GPa, σ_{max} = 610–690 MPa) (Perez-Rigueiro *et al.*, 2000) and their birefringence index was half that of native silk (Marsano *et al.*, 2005). However, the best processed fibers were more elastic (ε_{max} = 35%) (Marsano *et al.*, 2005) than the native silk (ε_{max} = 15%) (Perez-Rigueiro *et al.*, 2000).

Finally, a 10 wt% silk fibroin dope prepared with ionic liquids (1-ethyl-3-methylimidazolium chloride) was extruded into methanol. Wide-angle x-ray scattering (WAXS) data showed that the drawn fibers (DR = 2) had more oriented crystallites (Phillips *et al.*, 2005).

A patented process describes the first wet-spinning of fibers using recombinant spider silk proteins (Fahnestock, 1994). These silks had lower tensile strength and initial modulus (σ_{max} = 140 MPa, $E_{initial}$ = 4.6 MPa) than those reported for native or regenerated dragline silk (Seidel *et al.*, 2000).

A spider dragline silk cDNAs (ADF-3, or *Araneus diadematus* MaSp 2 protein) expressed in mammalian cells generated enough recombinant ADF-3 spider-silk proteins to spin fibers (Lazaris *et al.*, 2002). The 28% (w/v) spinning dope of the 60 kDa ADF-3 rc protein prepared in an aqueous buffer (160 mM urea, 10 mM Na_2HPO_4, 10 mM glycine, 1 mM tris pH 5) was extruded into 70–80% methanol. The double-drawn fibers (drawn in methanol and then water) had the best mechanical properties, though still not as strong or as tough as native dragline silk (Lazaris *et al.*, 2002).

Finally, a recent study reports the production, self-assembly characteristics, and wet-spinning of two synthetic spider silk-like analogs produced in *E. coli* (Teulé *et al.*, 2007). The basic repeats of these two proteins (60 kDa) contained a Flag-like elastic motif [(GPGGX$_1$ GPGGX$_2$)]$_2$, with X$_1$/X$_2$ equal to A/A or Y/S for the 'A1' or 'Y1' elastic versions respectively) adjacent to a MaSp 2-like strength motif {[linker-(poly(A)$_8$] called 'S8'}. The 'A1S8$_{20}$' and 'Y1S8$_{20}$' proteins were purified by nickel affinity chromatography after heat-treatment (at 80 °C for 10 min) of the total protein extracts. Circular dichroism (CD) data of melting (0–85 °C) and successive annealing (85–0 °C) for both proteins in aqueous buffers (5 mM Tris-HCl pH 8 or 0.1X PBS) showed a heat-inducible β-sheet transition irreversible only for Y1S8$_{20}$. Self-assembly of the pure proteins in aqueous environments was spontaneous for Y1S8$_{20}$ and shear-induced for A1S8$_{20}$. Moreover, 'dipping' forceps into the top layer of the pure protein fractions and pulling away resulted in fiber formation ('hand-pulled') in both cases (Fig. 3.3). Additionally, 25–30% (w/v) A1S8$_{20}$-HFIP and Y1S8$_{20}$-HFIP dopes extruded into 90% isopropyl alcohol generated fibers (Fig. 3.3). The fibers formed in aqueous environments had far better mechanical properties than those made from HFIP suggesting a better fiber internal organization. Within the 'pulled' fibers, the Y1S8$_{20}$ fibers were tougher (10.6 MJ.m^{-3}) and had an elasticity similar to that of dragline silk (ε_{max} = 34%)

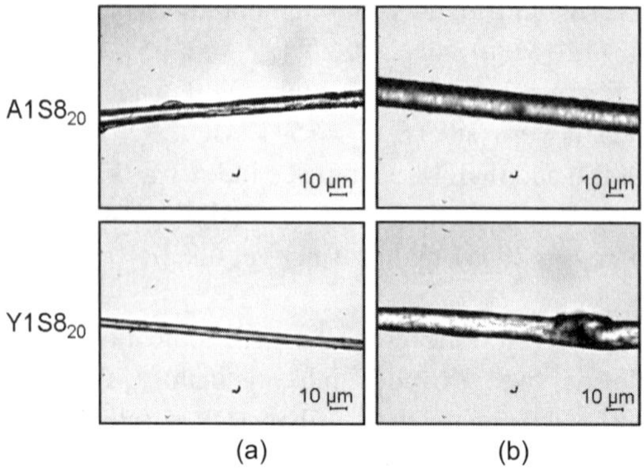

3.3 Synthetic spider silk artificial fibre. Light microscope observations of the A1S8$_{20}$ and Y1S8$_{20}$ artificial fibers (Teulé *et al.*, 2007): (a) fibers made from aqueous solutions ('pulled') and (b) fibers made from organic solvents ('as-spun'). Scale as indicated.

while displaying much lower maximum strength (σ_{max} = 50 MPa) (Teulé *et al.*, 2007) than dragline or flagelliform silks (σ_{max} = 4 GPa or 500 MPa) (Gosline *et al.*, 1999; Denny, 1976). However, within the HFIP-spun fibers, the A1S8$_{20}$ 'as-spun' fibers performed better than the Y1S8$_{20}$ ones therefore emphasizing the importance of fibrous protein primary structure and fiber formation conditions (Teulé *et al.*, 2007).

3.10 Electrospinning of fibrous proteins

3.10.1 What is 'electrospinning'?

In electrospinning, the polymer solution, supplied through a thin needle positioned opposite a collecting plate or target, is subjected to a high voltage (Fig. 3.4). Once the applied electric field overcomes the surface tension of the solution droplet, a jet forms and travels toward a grounded collecting plate. During this process, as a result of solvent evaporation, the jet thins down into nanoscale fibers that are deposited on the target plate as non-woven, highly porous, meshes (Bowlin, 2002). Using a rotating collector between the needle and the target plates (Xu *et al.*, 2004; Deitzel *et al.*, 2001) or a grounded mandrel (Matthews *et al.*, 2002) allows the collection of aligned fibers.

3.10.2 Electrospinning applied to the production of protein-based fibers

Electrospinning of collagen and elastin fibers

Initial reports suggested that the electrospinning of collagen from very acidic

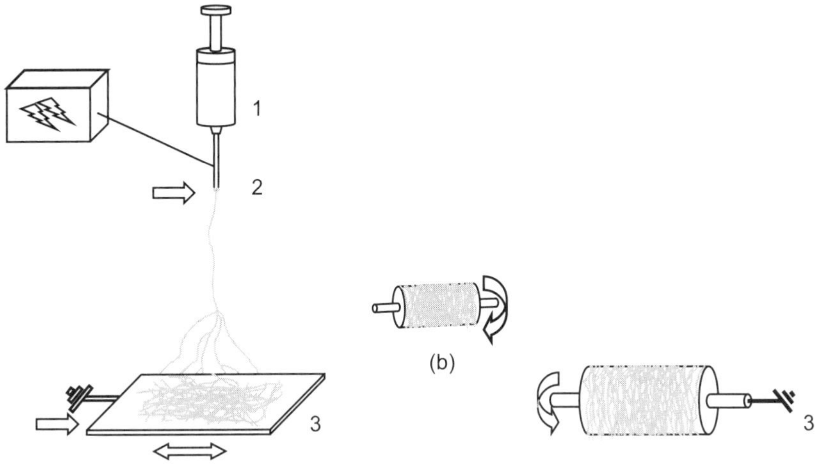

3.4 Electrospinning techniques: (a) example of a classic electrospinning set up. The spinning dope is pumped through a syringe (1) and exits through a capillary (2), which is connected to an adjustable high-voltage supply (box with lightning arrows). The formed polymer jet thins down, becomes unstable (splitting), and generates a mat of nanofibers collected on a grounded target/collection plate (3). For 'mixing electrospinning', the collection plate moves back and forth (double arrow) between two different polymer jets (second jet not represented) (Kidokaki *et al.*, 2005). (b) A rotating collector may be placed between the tip of the capillary and the target plate in (a) (Deitzel *et al.*, 2001; Xu *et al.*, 2004). (c) Alternatively, a rotating grounded mandrel may serve as a target in (a) (Matthews *et al.*, 2002). Both (b) and (c) allow the collection of more aligned fibers.

solutions such as 1–2 wt% type I collagen from rat tail tendons in HCl (pH 2) was not feasible (Huang *et al.*, 2001). However, increasing the viscosity of this solution with poly(ethylene oxide) (PEO) (such that collagen/PEO = 1:1 or 1:2) and its conductivity with NaCl (34 mM) allowed the non-toxic production of non-woven fiber networks by electrospinning (Huang *et al.*, 2001). The best conditions were when spinning a 2 wt% collagen/PEO (1:1) containing 34 mM NaCl at a flow rate of 100 μm min^{-1} under 18 kV (distance tip/target = 15 cm). Scanning electron microscopy (SEM) and transmission electron microscopy (TEM) pictures showed uniform and perfectly blended fibers with diameters of 100–150 nm thus demonstrating a good PEO/collagen interface. The dry fibers displayed poor mechanical properties (σ_{max} = 0.37 MPa, $E_{initial}$ = 12 MPa) (Huang *et al.*, 2001). A later study also reported the electrospinning of type I collagen from calf skin using the same method (Buttafoco *et al.*, 2006). The dope composed of 5% (w/v) of PEO/2% (w/v) collagen (PEO/collagen = 1:1)/42.4 mM NaCl was spun at a flow rate of 69 μl min^{-1} under 10–25 kV (20–30 cm) producing uniform collagen/PEO fibers with average diameters of 400 nm. The fibers became insoluble in water once crosslinked by treatment with 1-ethyl-3-dimethyl aminopropyl) carboiimide (EDC) in the presence of *N*-hydroxysuccinimide (NHS) in aqueous 70% ethanol. During this process, the PEO and NaCl totally leached out without disrupting the scaffold (Buttafoco *et al.*, 2006).

Organic solvents like HFIP or 2,2,2-trifluoroethanol (TFE) were also investigated. A first study reports the spinning of collagen isotypes from different tissue origins and the collection of aligned fibers using a rotating mandrel (Mathews *et al.*, 2002). The best conditions to generate calf skin type I collagen fibers were spinning a 8.3% (w/v) collagen–HFIP dope delivered at 83.4 μl/min in an electric field of 2 k•V cm^{-1} (25 kV). Aligned fibrils 100 nm in diameter were deposited on the surface of the mandrel moving at 1.4 m•s^{-1}. Sheets of spun material exhibited poor mechanical properties (σ_{max} = 1.3–1.7 MPa, $E_{initial}$ = 47.1–57.5 MPa). Under identical parameters, type I collagen from human placenta produced aligned filaments of slightly different structures with diameters of 100–730 nm. However, under these conditions, electrospinning human type III collagen required the use of 4% (w/v) collagen–HFIP dopes. Apparently, the source and collagen isotype and, thus, protein sequence, affect the polymer's structural properties (Mathews *et al.*, 2002). A different research group also spun calf skin type I collagen from a 10% (w/v) collagen–HFIP dope delivered at least at 83.34 μl min^{-1} and under 10 kV (15 cm) (Li *et al.*, 2005). In addition, an 8% (w/v) gelatin (bovine)-HFIP was spun using the same parameters. Both produced uniform collagenous mats with fibers of 200–500 nm. The gelatin and collagen fibers had tensile strengths of 8 and 12 MPa, respectively, and elasticities of 8 and 10%, respectively. The collagen fibers were much stiffer than the gelatin fibers ($E_{initial}$ = 262 versus 42.6 MPa) (Li *et al.*, 2005) and had improved mechanical properties compared to those published earlier (Matthews *et al.*, 2002).

A last study reports the electrospinning of type I collagen (bovine skin) and styrenated (ST) gelatin from HFIP using multilayering and mixing techniques (Kidokaki *et al.*, 2005). A 10 wt% ST–gelatin–HFIP dope was spun at a flow rate of 50 μl/min and using 25 kV (15 cm) whereas type I collagen was spun at the same flow rate from a 5 wt% dope and using 15 kV (15 cm). Both non-woven collagenous meshes contained fibers with diameters of 0.2–2 μm. To get insoluble fibers, the collagen fibers were crosslinked by UV light whereas the gelatin fibers were photopolymerized by visible light using camphorquinone. No mechanical data were reported (Kidokaki *et al.*, 2005). TFE was used to spin 5–12% (w/v) gelatin (porcine skin) solutions delivered at 13.4 μl/min under 10–12.5 kV (Huang *et al.*, 2004). The fiber mats collected contained fibers with average diameters of 100–340 nm. The best tensile strength (σ_{max} = 4.75 MPa) was achieved by fibers from mats spun using a 7.5% (w/v) gelatin–TFE dope. These fibers were also fairly elastic (ε_{max} = 6%) (Huang *et al.*, 2004).

The first example of successful electrospinning of elastin was achieved using 5–20 wt% solutions of an 81 kDa recombinant elastin solubilized in deionized water

(Huang *et al.*, 2000). Filaments of 0.2–3 μm composed the non-woven meshes which had interesting mechanical properties (σ_{max} = 35 MPa, $E_{initial}$ = 1.8 GPa) (Huang *et al.*, 2000).

Later, elastin was also electrospun using soluble bovine α-elastin and a 64 kDa recombinant human tropoelastin (Li *et al.*, 2005). Twenty per cent (w/v) α-elastin-HFIP or tropoelastin-HFIP (Martin *et al.*, 1995) dopes were spun at a flow rate of 16.7–50 μl/min using 10 kV (15 cm) (Li *et al.*, 2005). The fibers were wider and thicker (several micrometers) than those made from collagen or gelatin. The α-elastin fibers were the most brittle ($E_{initial}$ = 184 MPa, σ_{max} = 1.6 MPa, ε_{max} = 1%). The tropeoelastin fibers were as strong (σ_{max} = 13 MPa) as the collagen or gelatin fibers, more elastic (ε_{max} = 15%) than both collagenous fibers and as stiff ($E_{initial}$ = 289 MPa) as the collagen fibers (Li *et al.*, 2005).

Soluble bovine elastin was also spun from aqueous solutions composed of 5 wt% elastin/1% (w/v) PEO (elastin/PEO = 1:5)/42.5 mM NaCl delivered at 50 μl/min under 10 kV (25 cm) (Buttafoco *et al.*, 2006). The fibers had a rough surface and were 5–10 nm wide with a thickness of 0.5 μm. They had characteristics similar to those of native elastin. The elastin fibers became insoluble in water upon crosslinking and were then PEO- and NaCl-free (Buttafoco *et al.*, 2006). The same group also electrospun mixtures of collagen and elastin using PEO and NaCl. Collagen/elastin mixtures of 1–5% (w/w ratios of 2:1, 1:1, 1:2, or 1:3) containing 0.5 wt% PEO and 42.5 mM NaCl were spun at a flow rate of 30 μl/min under 10–30 kV (20–30 cm). The network was formed of perfectly blended and indistinguishable elastin and collagen fibers. The crosslinked insoluble collagen/elastin scaffolds supported the growth of smooth muscle cells (SMC) (Buttafoco *et al.*, 2006).

Electrospinning of silk fibers

In 2000, electrospinning allowed the regeneration of spider dragline (*N. clavipes*) and silkworm (*B. mori*) silk fibers with diameters of 100–200 nm and 6.5–100 nm, respectively (Reneker *et al.*, 2000). The patented process describes the spinning of 0.23–1.2 wt% dragline silk–HFIP and 0.74 wt% silk fibroin–HFIP solutions at 24–30 kV (15 cm). The fibers were annealed at different temperatures (Reneker *et al.*, 2000; Zarkoob *et al.*, 2004). TEM and wide-angle x-ray diffraction (WAXD) data indicated that crystal orientation only visible in the annealed fibers occurred at 230–280 °C and 205–240 °C for dragline and silkworm silks respectively. However, no mechanical data were reported (Zarkoob *et al.*, 2004). Another study reports the regeneration of silks from domesticated (*B. mori*) and wild (*Samia cynthia ricini*) silkworms through electrospinning using HFA-hydrate as a solvent (Ohgo *et al.*, 2003). These silks greatly differ in primary sequences and so far the wet-spinning of only *B. mori* fibroins using HFA-hydrate was successful (Yao *et al.*, 2002). The fibroin–HFA dopes prepared from *B. mori* and *S. c. ricini* silkworm silks (Ohgo *et al.*, 2003) were spun using an electric field of 1 kV/cm and concentrations of 3 and 10 wt%, respectively. Once dried, the fibers were immersed in methanol to initiate crystallization. During this process and after successive drying of the fibers under vacuum, the HFA-hydrate was completely removed. SEM data revealed that both types of regenerated fibers had diameters of 200–300 nm (Ohgo *et al.*, 2003). The ^{13}C crosspolarization (CP)/MAS NMR data showed a structural transition of the alanines from random coil (silk I) in the *B. mori* 'as-spun' fibers to anti-parallel β-sheet (silk II) caused by the methanol treatment. However, the data for the *S. c. ricini* fibers suggested that these structures formed upon drying the fibers, i.e. before methanol treatment. The mechanical properties of the regenerated *B. mori* (σ_{max} = 15 MPa, ε_{max} = 40%) and *S. c. ricini* (σ_{max} = 20 MPa, ε_{max} = 20%) fibers were different (Ohgo *et al.*, 2003).

Other research groups investigated the use of formic acid (98–100%) as a solvent to electrospin *B. mori* silk fibers using voltages of 10–50 kV (Sukigara *et al.*, 2003; Sukigara *et al.*, 2004; Ayutsede *et al.*, 2005; Min *et al.*, 2004). Initial experimental data showed that using 12–15 wt% silk fibroin–FA dopes and electric fields of 3–4 k/cm regenerated uniform fibers with diameters smaller than 100 nm (Sukigara *et al.*, 2003). Later, when a response surface methodology (RSM) approach was applied to the experimental data, a model predicted that electrospinning 8–10 wt% silk fibroin–FA dopes with electric fields of 4–5 kV/cm should produce fibers of diameters smaller than 40 nm (Sukigara *et al.*, 2004). Electrospinning 9–15 wt% silk fibroin–FA dopes at electric fields of 2–4 kV/cm (10–50 kV) generated fibers with circular diameters smaller than 100 nm (Ayutsede *et al.*, 2005). FTIR, Raman spectroscopy and WAXD data indicated that the regenerated fibers were more crystalline (48%) than native silks (39%). The authors observed that the dissolution of fibroins in FA enhanced β-sheet crystallization and may facilitate β-sheet formation in the electrospun fiber. However, according to WAXD data, the lack of crystallite orientation in the 'as-spun' fiber was ultimately responsible for the poor mechanical properties of the fiber mat produced ($E_{initial}$ = 515 MPa, σ_{max} = 7.25 MPa, ε_{max} = 3.2%) (Ayutsede *et al.*, 2005) compared with that of the native fiber ($E_{initial}$ = 15–17 GPa, σ_{max} = 610–690 MPa, ε_{max} = 15%) (Perez-Rigueiro *et al.*, 2000). Note that no post-spinning processing was done on these fibers (Ayutsede *et al.*, 2005). A last study reports the electrospinning of 3–5 wt% silk–fibroin–FA dopes at 15 kV (7 cm) to regenerate silkworm silk fibers (Min *et al.*, 2004). The fibers were successively treated with 50% methanol. SEM data showed circular and smooth fibers with an average diameter of 80 nm, and a wide range of pore sizes desirable for cell attachment. Though no mechanical performance was reported, the fiber mats promoted cell adhesion and spreading of type I collagen (Min *et al.*, 2004).

Finally, a different study provides an alternative protocol to electrospin nanometer-scale protein fibers from aqueous silk fibroin dopes using PEO as an additive (Jin *et al.*, 2002). Silk fibroin/PEO dopes of concentrations from 4.8–8.8 wt% spun at flow rates of 20–50 μl/ml using electric fields between 0.5–0.6 kV/cm (10–12 kV) generated fibers with diameters averaging 800 nm. FTIR data showed that silk I was the predominant structure in the 'as-spun' fibers and that successive immersion of the fiber mats in aqueous 90% methanol for 10 min induced β-sheet formation (Jin *et al.*, 2002).

Very few studies report the electrospinning of recombinant silk-like proteins to produce fibers. Previous work described the production of the genetically engineered SLPF (silk-like polymer with fibronectin) hybrid protein characterized by the crystalline $(GAGAGS)_9$ segment from *B. mori* silk adjacent to a fibronectin sequence containing the RGD tripeptide promoting cell attachment (Anderson *et al.*, 1994). This SFPL polymer was dissolved in 96% FA to prepare 0.8–16.2 wt% dopes that were spun using electric fields of 2–8 kV/cm (Buchko *et al.*, 1999). Using concentrations equal or superior to 12.1 wt% resulted in the formation of non-beaded filamentous coatings. Only the SLPF films were later tested and mechanical data showed that they were relatively brittle ($\varepsilon_{max} = 3\%$) (Buchko *et al.*, 2000). Later, in a different study, a recombinant silkworm silk protein composed of adjacent *B. mori* crystalline and *S. c. ricini* glycine-rich segments was used to prepare a chimeric silkworm silk–HFA dope that was successfully electrospun (Ohgo *et al.*, 2003). SEM observations showed homogenous chimeric silkworm silk fibers with diameters of 100 nm though no mechanical data were obtained.

3.11 Applications

Such synthetic protein fibers with customized mechanical properties represent important potential alternatives as more natural materials for several medical, civil and military applications. In the medical field, provided they are biocompatible, these protein-based fibers could be used as strong suture threads, or artificial ligaments. Additionally, the architecture of the produced electrospun meshes which naturally resemble that found in most extracellular matrices make these materials suited for use as protein-based scaffolds for tissue and bone regeneration. Ropes, fabrics or filtering systems are some examples of possible civilian uses while antiballistic gear, strong light-weight gear, parachutes and harnesses constitute possible military applications.

3.12 Future trends and conclusions

Studies show that the spinning of native or synthetic fibrous protein materials is feasible and may soon generate a new source of natural materials on the market. However, to achieve this goal, the spinning technologies adapted to the creation of customized high-performance protein fibers need to be further developed. This may require a more complete knowledge of the behavior of the individual protein materials used in solutions and deeper investigations of the necessary protein structural transitions or refolding processes that are crucial to controlling proper self-assembly. Cleverly designed synthetic protein analogs can be produced to help determine the molecular mechanisms that confer strength and/or elasticity.

However, further investigation is needed to optimize the primary structures of model fibrous proteins as well as the processing of the produced synthetic analogs. Moreover, since these protein-based fibers are designed to be 'more natural' materials, the ultimate goal is to produce synthetic fibers using innocuous processes. Therefore, research efforts in fiber production may shift mostly toward the use of aqueous non-toxic and non-polluting solvents instead of harsh organic solvents. Non-toxic recyclable organic solvents such as NMMO still seem to be appropriate for the production of synthetic protein fibers as demonstrated in the case of regenerated silk fibers (Marsano *et al.*, 2005).

The use of non-toxic, ultimately removable, chemical polymers such as PEO in blends with fibrous proteins has proven to be a way to produce fibers in aqueous conditions (Jin *et al.*, 2002; Buttafoco *et al.*, 2006). Although synthetic protein fibers have been generated, their mechanical performances need to be improved to become viable products. An alternative way to modify the mechanical and chemical properties of synthetic protein fibers is to blend fibrous protein materials with other known natural non-protein fibrous materials such as polysaccharides. Examples of such biofiber blends are the production of synthetic 'natural' fiber blends of silk fibroin/chitin (Park *et al.*, 2006) or fibroin/cellulose and fibroin/chitin/cellulose (Hirano *et al.*, 2002). In another case, the blending of collagen with chitosan allowed the production of fibers with improved blood biocompatibility thus allowing their possible use in wound dressing treatments (Hirano *et al.*, 2000). Recently, a novel chimeric fusion protein containing silk and silica-forming domains was used to create new silk–silica films and nanoscale protein fibers in aqueous conditions thus opening the door to a new kind of interesting biomaterial (Wong Po Foo *et al.*, 2006b). All of these aspects of natural protein fiber production are still currently under active research. This ever-expanding field offers exciting and promising possibilities for the creation and exploitation of more natural yet smarter biological materials.

3.13 References

Akai H (1983), 'The structure and ultrastructure of the silk gland', *Experientia*, **39**, 443–449.

Andersen SO (1964), 'The crosslinks in resilin identified as dityrosine and trityrosine', *Biochim Biophys Acta*, **93**, 213–215.

Anderson JP, Cappello J and Martin DC (1994), 'Morphology and primary crystal structure of a silk-like protein polymer synthesized by genetically engineered *Escherichia coli* bacteria', *Biopolymers*, **34**, 1049–1057.

Asakura T, Umema K, Nakazawa Y, Hirose H, Higham J and Knight D (2007), 'Some observations on the structure and function of the spinning apparatus in the silkworm *Bombyx mori*', *Biomacromolecules*, **8**, 175–181.

Ayutsede J, Gandhi M, Sukigara S, Micklus M, Chen H-E and Ko F (2005), 'Regeneration of *Bombyx mori* silk by electrospinning. Part 3: characterization of electrospun nonwoven mat', *Polymer*, **46**, 1625–1634.

Bochicchio B, Pepe A and Tamburro AM (2001), 'On (GGLGY) synthetic repeating sequences of lamprin and analogous sequences', *Matrix Biol*, **20**, 243–250.

Bowlin GL (2002), 'Electrospinning collagen scaffolds', *Sci Med*, **8**, 292–293.

Buchko CJ, Chen LC, Shen Y and Martin DC (1999), 'Processing and microstructural characterization of porous biocompatible protein polymer thin films', *Polymer*, **40**, 7397–7407.

Buchko CJ, Slattery MJ, Kozloff KM and Martin DC (2000), 'Mechanical properties of biocompatible protein polymer thin films', *J. Mater Res*, **15**(1), 231–242.

Buttafoco L, Kolkman NG, Engbers-Buijtenhuijs P, Poot AA, Dijkstra PJ, Vermes I and Feijen J (2006), 'Electrospinning of collagen and elastin for tissue engineering applications', *Biomaterials*, **27**, 724–734.

Chapman M (1969), *Text Res J*, **39**, 1102.

Chen X, Knight DP, Shao Z and Vollrath F (2002), 'Conformation transition in silk protein films monitored by Time Resolved Fourier Transform Infrared Spectroscopy. Effect of potassium ions on *Nephila* spidroin films', *Biochemistry*, **41**, 14944–14950.

Colgin MA and Lewis RV (1998), 'Spider minor ampullate silk proteins contain new repetitive sequences and highly conserved non silk-like "spacer regions" ', *Protein Sci*, **7**, 667–672.

Cox GN (1992), 'Molecular and biochemical aspects of nematode collagens', *J Parasitol*, **78**, 1–15.

Coyne KJ, Qin XX and Waite JH (1997), 'Extensible collagen in mussel byssus. A natural block copolymer', *Science*, **277**, 1830–1832.

Coyne KJ and Waite JH (2000), 'In search of molecular dovetails in mussel byssus: from the threads to the stem', *J Exp Biol*, **198**, 633–644.

Craig CL (1997), 'Evolution of arthropod silks', *Annu Rev Entomol*, **42**, 231–267.

Datta A, Ghosh AK and Kundu SC (2001), 'Differential expression of fibroin gene in development stages of silkworm *A mylitta (Saturniidae)*', *Comp Biochem Physiol B Biochem Mol Biol*, **129**(1), 197–204.

Debelle L and Alix AJP (1999), 'The structure of elastins and their function', *Biochimie*, **81**(10), 981–984.

Deitzel JM, Kleinmeyer J, Harris D and Tan NC (2001), 'The effect of processing variables on the morphology of electrospun nanofibers and textiles', *Polymer*, **42**(1), 261–272.

Denny MW (1976), 'The physical properties of spider silks and their role in the design of orb webs', *J Exp Biol*, **65**, 483–505.

Denny MW (1980), 'Silks—their properties and functions' in 'The mechanical properties of biological materials', *Symp Soc Exp Biol*, **34**, Cambridge, Cambridge University Press, 247–272.

Dicko C, Vollrath F and Kenney JM (2004a), 'Spider silk protein refolding is controlled by changing pH', *Biomacromolecules*, **5**, 704–710.

Dicko C, Vollrath F and Kenney JM (2004b), 'Secondary structures and conformational changes in flagelliform, cylindrical, major, and minor ampullate silk proteins. Temperature and concentration effects', *Biomacromolecules*, **5**, 2105–2115.

Dicko C, Kenney JM, Knight D and Vollrath F (2004c), 'Transition to a β-sheet-rich structure in spidroin *in vitro*: the effect of pH and cations', *Biochemistry*, **43**, 14080–14087.

Edens WA, Sharling L, Cheng G, Shapira R, Kinkade J, Lee T, Edens HA, Tang X, Sullard C, Flaherty DB, Benian GM and Lambeth JD (2001), 'Tyrosine cross-linking of extracellular matrix is catalyzed by Duox, a multidomain oxidase/peroxidase with homology to the phagocyte oxidase subunit gp91phox', *J Cell Biol*, **154**(4), 879–891.

Fahnestock SL (1994), 'Novel, recombinantly produced spider silk analogs', Du Pont De Nemours and Company, European Patent No 1413585.

Feughelman M (2000), 'Natural protein fibers', *J Appl Polym Sci*, **83**, 489–507.

Foelix RF (1996), '*Biology of spiders*', Second edition, Oxford University Press Georg Thieme Verlag.

Fofonoff TW and Bell E (1999), 'Method for spinning and processing collagen fiber', US Patent No 5911942.

Fukae R, Maekawa A and Sangen O (2005), 'Gel-spinning and drawing of gelatin', *Polymer,* **46**, 11193–11194.

Furukawa M, Takada M, Shoichi M and Sasayama A (1994), 'Process for producing regenerated collagen fiber', US Patent No 5344917.

Gatesy J, Hayashi C, Motriuk D, Woods J and Lewis R (2001), 'Extreme diversity, conservation, and convergence of spider silk fibroin sequences', *Science*, **291**(3), 2603–2605.

Gosline JM (1978), 'Hydrophobic interaction and a model for the elasticity of elastin', *Biopolymers,* **17**, 677–695.

Gosline JM, Denny M and DeMont M (1984), 'Spider silk as a rubber', *Nature*, **309**, 551–552.

Gosline JM (1987), 'Structure and mechanical properties of rubber-like proteins in animals', *Rubber Chem Technol,* **60**, 417–438.

Gosline J, Guerette P, Ortlepp C and Savage K (1999), 'The mechanical design of spider silks: from fibroin sequence to mechanical function', *J Exp Biol*, **202**, 3295–3303.

Grzelak K (1995), 'Control of expression of silk protein genes', *Comp Biochem Physiol B*, **110**, 671–681.

Guerette PA, Ginzinger DG, Weber BHF and Gosline JM (1996), 'Silk properties determined by gland-specific expression of a spider fibroin gene family', *Science*, **272**, 112–115.

Ha S-W, Tonelli AE and Hudson SM (2005), 'Structural studies of *Bombyx mori* silk fibroin during regeneration from solutions and wet fiber spinning', *Biomacromolecules*, **6**, 1722–1731.

Hayashi CY and Lewis RV (1998), 'Evidence from flagelliform silk cDNA for the structural basis of elasticity and modular nature of spider silks', *Mol Biol*, **275**, 773–784.

Hayashi CY, Shipley NH and Lewis RV (1999), 'Hypotheses that correlate the sequence, structure, and mechanical properties of spider silk proteins', *Int J Biol Macrom*, **24**, 271–275.

Hayashi CY and Lewis RV (2000), 'Molecular architecture and evolution of a modular spider silk protein gene', *Science*, **287**, 1477–1479.

Hinman M and Lewis RV (1992), 'Isolation of a clone encoding a second dragline silk fibroin', *J Biol Chem*, **267**(27), 19320–19324.

Hirano S, Zhang M, Nakawaga M and Miyata T (2000), 'Wet-spun chitosan-collagen fibers, their chemical N-modifications, and blood compatibility', *Biomaterials*, **21**, 997–1003.

Hirano S, Nakahira T, Zhang M, Nakagawa M, Yoshikawa M and Midorikawa T (2002), 'Wet-spun blend biofibers of cellulose-silk fibroin and cellulose-chitin-silk fibroin', *Carbohydr Polym,* **47**, 121–124.

Hronska M, van Beek JD, Williamson PTF, Vollrath F and Meier BH (2004), 'NMR characterization of native liquid spider dragline silk from *Nephila edulis*', *Biomacromolecules*, **5**, 834–839.

Huang L, McMillan RA, Apkarian RP, Pourdeyhimi B, Conticello VP and Chaikof EL (2000), 'Generation of synthetic elastin-mimetic small diameter fibers and fiber networks', *Macromolecules*, **33**, 2989–2997.

Huang L, Nagapudi K, Apkarian RP and Chaikof EL (2001), 'Engineered collagen-PEO nanofibers and fabrics', *J Biomater Sci Polym Ed*, **12**(9), 979–993.

Huang ZM, Zhang YZ, Ramakrishna S and Lim CT (2004), 'Electrospinning and mechanical characterization of gelatin nanofibers', *Polymer*, **45**, 5361–5368.

Indik Z, Yeh H, Ornsten-Goldstein N, Shepard P, Anderson N, Rosembloom JC, Peltonen L and Rosembloom J (1987), 'Alternative splicing of human elastin mRNA indicated by sequence analysis of cloned genomic and complementary DNA', *Proc Natl Acad Sci USA*, **84**, 5680–5684.

Jelinski LW, Blye A, Liivak O, Michal C, La Verde G, Seidel A, Shah N and Yang Z (1999), 'Orientation, structure, wet-spinning, and molecular basis for supercontraction of spider dragline silk', *Int J Biol Macromol*, **24**, 197–201.

Jin H-J, Fridrikh SV, Rutledge GC and Kaplan DL (2002), 'Electrospinning *Bombyx mori* silk with poly(ethylene oxide)', *Biomacromolecules*, **3**, 1233–1239.

Kadler KE, Holmes DF, Trotter JA and Chapman JA (1996), 'Collagen fibril formation', *Biochem J*, **316**, 1–11.

Kerkam K, Viney C, Kaplan D and Lombardi S (1991), 'Liquid crystallinity of natural silk secretions', *Nature,* **349**, 596–598.

Kidokaki S, Kwon IK and Matsuda T (2005), 'Mesoscopic spatial designs of nano- and microfiber meshes for tissue engineering matrix and scaffold based on newly devised multilayering and mixing electrospinning techniques', *Biomaterials*, **26**, 37–46.

Kikuchi K, Mori K, Suzuki Y, Yamaguchi K and Mizuno S (1992), 'Structure of the *Bombyx mori* fibroin light-chain encoding gene: upstream sequence elements common to the light and heavy chains', *Gene*, **110**, 151–158.

Knight DP and Vollrath F (1999), 'Liquid crystals and flow elongation in a spider's silk production line', *Proc R Soc Lond B*, **266**, 519–523.

Knight DP, Knight MM and Vollrath F (2000), 'Beta-transition and stress-induced phase separation in the spinning of spider dragline silk', *Int J Biol Macromol*, **27**, 205–210.

Knight DP and Vollrath F (2001), 'Changes in element composition along the spinning duct in a *Nephila spider*', *Naturwissenschaften*, **88**, 179–182.

Kovoor J (1977), 'La soie et les glandes séricigènes des arachnides', *Ann Biol*, **16**, 97–171.

Kovoor, J (1986), 'Chapter IV – comparative structure and histochemistry of silk producing organs in arachnids', *in 'Ecophysiology of spiders'*, edited by Nentwig W, Springer Verlag, Berlin/Heidelberg/New York, 160–186.

Kovoor J (1990), 'The silk gland system in some *Tetragnatinae* (*Aranea: Araneidae*). Comparative anatomy and histochemistry', *Acta Zool Fenn*, **190**, 215–222.

Kramer JM (1994), 'Structures and functions of collagens in *Caenorhabditis elegans*', *FASEB*, **8**, 329–336.

Kusuda J, Tazima Y, Onimaru K, Ninaki O and Suzuki Y (1986), 'The sequence around the 5' end of the fibroin gene from the wild silkworm, *Bombyx mandarina*, and comparison with that of domesticated species *B mori*', *Mol Gen Genet*, **203**(3), 359–364.

Lazaris A, Arcidiacono S, Huang Y, Zhou J-F, Duguay F, Chretien N, Welsh EA, Soares JW and Karatzas CN (2002), 'Spider silk fibers spun from soluble recombinant silk produced in mammalian cells', *Science*, **295**, 472–476.

Lewis RV (1992), 'Spider silk: the unraveling of a mystery', *Acc Chem Res*, **25**, 382–398.

Lewis RV, Hinman M, Kothakota S and Fournier MJ (1996), 'Expression and purification of a spider silk protein: a new strategy for producing repetitive proteins', *Protein Expr Purif*, **7**, 400–406.

Li M, Mondrinos MJ, Gandhi MR, Ko FK, Weiss AS and Lelkes PI (2005), 'Electrospun protein fibers as matrices for tissue engineering', *Biomaterials*, **26**, 5999–6008.

Liivak O, Blye A, Shah N and Jelinski LW (1998), 'A microfabricated wet-spinning apparatus to spin fibers of silk proteins–structure–property correlations', *Macromolecules*, **31**, 2947–2951.

Lock RL (1993), 'Process for making silk fibroin fibers', Du Pont De Nemours and Company, US Patent No 5252285.

Lucas F (1964), 'Spiders and their silks', *Discovery*, **25**(1), 20–26.

Magoshi J, Magoshi Y and Nakamura S (1985a), 'Crystallization, liquid crystal, and fiber formation of silk fibroin', *J Appl Polym Sci*, **41**, 187–204.

Magoshi J, Magoshi Y and Nakamura S (1985b), 'Physical properties and structure of silk: The mechanism of fibre formation from liquid silk of silkworm *Bombyx mori*', *Polym Commun*, **26**, 165–177.

Marsano E, Corsini P, Arosio C, Boschi A, Mormino M and Freddi G (2005), 'Wet spinning of *Bombyx mori* silk fibroin dissolved in *N*-methylmorpholine *N*-oxide and properties of the regenerated fibres', *Int J Biol Macromol*, **37**, 179–188.

Martin SL, Vrhovski B and Weiss AS (1995), 'Total synthesis and expression in *Escherichia coli* of a gene encoding human tropoelastin', *Gene*, **154**, 159–166.

Mathews JA, Wnek GE, Simpson DG and Bowlin GL (2002), 'Electrospinning of collagen nanofibers', *Biomacromolecules*, **3**, 232–238.

Min B-M, Lee G, Kim SH, Nam YS, Lee TS and Park WH (2004), 'Electrospinning of silk fibroin nanofibers and its effect on the adhesion and spreading of normal human keratinocytes and fibroblasts *in vitro*', *Biomaterials*, **25**, 1289–1297.

Mita K, Ishimura S and James TC (1994), 'Highly repetitive structure and its organization of the silk fibroin gene', *J Mol Evol*, **38**(6), 583–592.

Nagura M, Yokota H, Ikeura M, Gotoh Y and Ohkoshi Y (2002), 'Structures and physical properties of cross-linked gelatin fibers', *Polym J*, **34**, 761–766.

Nakamae K, Nishino T and Ohkubo H (1989), 'Elastic modulus of the crystalline regions of silk fibroin', *Polymer*, **30**, 1243–1246.

Ohgo K, Zhao C, Kobayashi M and Asakura T (2003), 'Preparation of non-woven nanofibers of *Bombyx mori*, *Samia cynthia ricini* silk and recombinant hybrid silk with electrospinning method', *Polymer*, **44**, 841–846.

Park KE, Jung SY, Lee SJ, Min B-M and Park WH (2006), 'Biomimetic nanofibrous scaffolds: preparation and characterization of chitin/silk fibroin blend nanofibers', *Int J Biol Macromol*, **38**, 165–173.

Perez-Rigueiro J, Viney C, Llorca J and Elices M (2000), 'Mechanical properties of single-brin silkworm silk', *J Appl Polym Sci*, **75**(10), 1270–1277.

Phillips DM, Drummy LF, Naik RR, De Long HC, Fox DM, Trulove PC and Mantz RA (2005), 'Regenerated silk fiber wet spinning from an ionic liquid solution', *J Mater Chem*, **15**, 4206–4208.

Pollock CM and Shadwick RE (1994), 'Relationship between body mass and biomechanical properties of limb tendons of adult mammals', *Regul Integrative Comp Physiol*, **35**, 1016–1021.

Pujol JP (1970), 'Le collagène du byssus de *Mytilus edulis* L', *Z Zellforsch*, **104**, 358–374.

Qin XX and Waite JH (1998), 'A potential mediator of collagenous block copolymer gradients in mussel byssal threads', *Proc Natl Acad Sci USA*, **95**, 10517–10522.

Qin XX, Coyne KJ and Waite JH (1995), 'Exotic collagen gradients in the byssus of the mussel *Mytilus edulis* L', *J Exp Biol*, **198**, 633–644.

Qin XX, Coyne KJ and Waite JH (1997), 'Tough tendons. Mussel byssus has collagen with silk-like tendons', *J Biol Chem*, **272**(51), 32623–32627.

Rausher S, Baud S, Miao M, Keeley FW and Pomès R (2006), 'Proline and glycine control protein self-organization into elastomeric or amyloid fibrils', *Structure*, **14**, 1667–1676.

Ray C, Wang TY and Hussey RS (1996), 'Identification and characterization of the *Meloidogyne incognita col*-1 cuticle collagen gene', *Mol Biochem Parasitol*, **83**, 121–124.

Reneker DH, Eby RK, Ertley D, Hudson SD and Zarkoob S (2000), 'Synthetically spun silk nanofibers and a process for making the same', University of Akron, US Patent No 6110590.

Robson P, Wright GM, Sitarz E, Maiti A, Rawa M, Youson JH and Keeley FW (1993), 'Characterization of lamprin, an unusual matrix protein from lamprey cartilage. Implications for evolution, structure, and assembly of elastin and other fibrillar proteins', *J Biol Chem*, **268**(2), 1440–1447.

Sehnal F and Akai H (1990), 'Insect silk glands: their types, development and function, and effects of environmental factors and morphogenetic hormones on them', *Int J Insect Morphol Embryol*, **19**, 79–132.

Seidel A, Liivak O and Jelinski LW (1998), 'Artificial spinning of spider silk', *Macromolecules*, **31**, 6733–6736.

Seidel A, Liivak O, Calve S, Adaska J, Ji G, Yang Z, Grubb D, Zax DB and Jelinski LW (2000), 'Regenerated spider silk: processing, properties and structure', *Macromolecules*, **33**, 775–780.

Shewry PR, Halford NG and Tatham AS (1992), 'The high molecular weight subunits of wheat glutenin', *J Cereal Sci*, **15**, 105–120.

Sponner A, Schlott B, Vollrath F, Unger E, Grosse F and Weisshart K (2005), 'Characterization of the protein components of *Nephila clavipes* dragline silk', *Biochemistry*, **44**, 4727–4736.

Stauffer S, Coguill S and Lewis R (1994), 'Comparison of physical properties of three silks from *Nephila clavipes* and *Araneus gemmoides*', *J Arachnol*, **22**, 5–11.

Sukigara S, Gandhi M, Ayutsede J, Micklus M and Ko F (2003), 'Regeneration of *Bombyx mori* silk by electrospinning. Part 1: processing parameters and geometric properties', *Polymer*, **44**, 5721–5727.

Sukigara S, Gandhi M, Ayutsede J, Micklus M and Ko F (2004), 'Regeneration of *Bombyx mori* silk by electrospinning. Part 2: Process optimization and empirical modeling using response surface methodology', *Polymer*, **45**, 3701–3708.

Tanaka K, Inoue S and Mizuno S (1999a), 'Hydrophobic interactions of P25, containing Asn-linked oligosaccharide chains, with the H-L complex of silk fibroin produced by *Bombyx mori*', *Insect Biochem Mol Biol*, **29**, 269–276.

Tanaka K, Kajiyama N, Ishikura K, Waga S, Kikuchi A, Ohtomo K, Takagi T and Mizuno S (1999b), 'Determination of the site of disulfite linkage between heavy and light chains of silk fibroin produced by *Bombyx mori*', *Biochim Biophys Acta*, **1432**(1), 92–103.

Tatham AS and Shewry PR (2000), 'Elastomeric proteins: biological roles, structures and mechanisms', *Trends Biol Sci*, **299**, 567–571.

Tatham AS and Shewry PR (2002), 'Comparative structure and properties of elastic proteins', *Phil Trans R Soc Lond B*, **357**, 229–234.

Teulé F, Furin WA, Cooper AR, Duncan JR and Lewis RV (2007), 'Modifications of spider silk sequences in an attempt to control the mechanical properties of the synthetic fibers', *J Mater Sci*, **42**, 8974–8985.

Tian MZ and Lewis RV (2004), 'Analysis of major ampullate silk cDNAs from two non-orb weaving spiders', *Biomacromolecules*, **5**(3), 657–661.

Tnau Coimbatore (2003), 'PH Sericulture—Lesson 8: Morphology and biology of silkworm', *http://www.tnau.ac.in/notesbscag/ser*, accessed November 2006.

Trabbic KA and Yager P (1998), 'Comparative structural characterization of naturally- and synthetically-spun fibers from *Bombyx mori* fibroin', *Macromolecules*, **31**, 462–471.

Tskhovrebova L and Trinick J (2003), 'Titin: properties and family relationships', *Nature Rev Mol Cell Biol*, **4**, 679–689.

Um IC, Ki CS, Kweon HY, Lee KG, Ihm DW and Park YH (2004), 'Wet spinning of silk polymer. II. Effect of drawing on the structural characteristics and properties of filament', *Int J Biol Macromol*, **34**, 107–109.

Urry DW, Harris R D, Long MM and Prasad KU (1986), 'Polytetrapeptide of elastin: temperature correlated elastomeric force and structure development', *Int J Peptide Protein Res,* **28**, 649–660.

Urry DW (1988), 'Entropic elastic processes in protein mechanisms. 1. Elastic structures due to an inverse temperature transition and elasticity due to internal chain dynamics', *J Protein Chem,* **7**, 1–34.

Ushiki T (2002), 'Collagen fibers, reticular fibers and elastic fibers. A comprehensive understanding from a morphological viewpoint', *Arch Histol Cytol*, **65**(2), 109–126.

VanderEycken W, deAlmeidaEngler J, VanMontagu M and Gheysen G (1994), 'Identification and analysis of a cuticular collagen-encoding gene from the plant-parasitic nematode *Meloidogyne incognita*', *Gene*, **151**, 237–242.

Viney C, Huber AE, Dunaway DL, Kerkam K and Case ST (1994), 'Optical characterization of silk secretions and fibers' in '*Silk polymers—materials science and biotechnology*', edited by Kaplan DL, Adams WW, Farmer BL, Viney C, Washington DC, American Chemical Society.

Viney C (1997), 'Liquid crystalline phase behavior of proteins and polypeptides' in '*Protein-based materials*', edited by McGrath KC and Kaplan D, Birkhaüser, Boston, 281–311.

Vollrath F and Tillinghast EK (1991), 'Glycoprotein glue beneath a spiders web's aqueous coat', *Naturwissenschaften,* **78**, 557–559.

Vollrath F, Wen Hu X and Knight DP (1998), 'Silk production in a spider involves acid bath treatment', *Proc R Soc B*, **263**, 817–820.

Vollrath F and Knight D (1999), 'Structure and function of the silk production pathway in the spider *Nephila edulis*', *Int J Biol Macromol*, **24**, 243–249.

Vollrath F and Knight DP (2001), 'Liquid crystalline spinning of spider silks', *Nature*, **410**, 541–548.

Waite JH (1992), 'The formation of mussel byssus: anatomy of a natural manufacturing process', *Results Problems Cell Differentiation*, **19**, 27–54.

Waite JH, Qin XX, Coyne KJ (1998), 'The peculiar collagens of mussel byssus', *Matrix Biol*, **17**, 93–106.

Wallace D and Thompson A (1983), 'Description of collagen fibril formation by a theory of polymer crystallization', *Biopolymers*, **25**, 1783–1811.

Wang T, Deom CM and Hussey RS (1998), 'Identification of a *Meloidogyne incognita* cuticle collagen gene and characterization of the developmental expression of three collagen genes in parasitic stages', *Mol Biocheml Parasitol*, **93**, 131–134.

Weis-Fogh T (1960), 'A rubber-like protein in insect cuticle', *J Exper Biol*, **37**, 887–907.

Weis-Fogh T (1961), 'Molecular interpretation of the elasticity of resilin, a rubber-like protein', *J Mol Biol*, **3**, 520–531.

Weiskopf A, Senecal K, Vouros P, Kaplan D and Mello CM (1996), 'The carbohydrate composition of spider silk: *Nephila edulis* dragline', *Glycobiology*, **6**, 1703–1708.

Willcox PJ, Gido SP, Muller W and Kaplan DL (1996), 'Evidence of a cholesteric liquid crystalline phase in natural silk spinning processes', *Macromolecules,* **29**, 5106–5110.

Wilson R, Lees JF and Bulleid NJ (1998), 'Protein disulphide isomerase acts as a molecular chaperone during the assembly of procollagen', *J Biol Chem*, **273**, 9637–9643.

Winkler S, Szela S, Avtges P, Valluzzi R, Kirschner DA and Kaplan D (1999), 'Designing recombinant spider silk proteins to control assembly', *Int J Biol Macromol*, **24**, 265–270.

Winkler S, Wilson D and Kaplan D (2000), 'Controlling β-sheet assembly in genetically engineered silk by enzymatic phosphorylation/dephosphorylation', *Biochemistry*, **39**, 12739–12746.

Wong Po Foo C, Bini E, Hensman J, Knight DP, Lewis RV and Kaplan DL (2006a), 'Role of pH and charge on silk protein assembly in insects and spiders', *Appl Phys A*, **82**, 223–233.

Wong Po Foo C, Bini E, Huang J, Lee SY and Kaplan DL (2006b), 'Solution behavior of synthetic silk peptides and modified recombinant silk proteins', *Appl Phys A*, **82**, 193–203.

Xu M and Lewis RV (1990), 'Structure of a protein superfiber: spider dragline silk', *Proc Natl Acad Sci USA*, **87**, 7120–7124.

Xu CY, Inai R, Kotaki M and Ramakrishna S (2004), 'Aligned biodegradable nanofibrous structure: a potential scaffold for blood vessel engineering', *Biomaterials*, **25**(5), 877–886.

Yao J, Masuda H, Zhao C and Asakura T (2002), 'Artificial spinning and characterization of silk fiber from *Bombyx mori* silk fibroin in hexafluoroacetone hydrate', *Macromolecules*, **35**, 6–9.

Zarkoob S, Eby RK, Reneker DH, Hudson SD, Ertley D and Adams WW (2004), 'Structure and morphology of electrospun silk nanofibers', *Polymer*, **45**, 3973–3977.

Zhou CZ, Confalonieri F, Medina N, Zivanovic Y, Esnault C, Yang T, Jacquet M, Janin J, Duguet M, Perasso R and Li ZG (2000), 'Fine organization of *Bombyx mori* fibroin heavy chain gene', *Nucleic Acids Res*, **28**(12), 2413–2419.

Zhou L, Chen X, Shao Z, Zhou P, Knight DP and Vollrath F (2003), 'Copper in the silk formation process of *Bombyx mori* silkworm', *FEBS Lett*, **554**(3), 337–341.

4

Biomimetic principles of spider silk for high-performance fibres

C. HOLLAND and F. VOLLRATH
University of Oxford, UK

Abstract: The desirable properties of silk and the relationships between structure and composition are explored. Fibres spun from feedstocks with identical rheology have differing mechanical characteristics. However, supercontraction has linked the role of an individual amino acid, proline, to the mechanical properties of dragline silks. The practical difficulties in matching feedstock chemistry with the appropriate extrusion/spinning conditions for copying native silks, spider and silkworm, are discussed. Both processing conditions and protein composition are important. Producing commercial silk the spider's way would give excellent fibres through an ecologically compatible process.

Key words: silk, spider, silkworm, supercontraction, rheology.

4.1 Introduction

In order to mimic or copy silk we must first understand it. Understanding means not only knowing the relevant protein motifs but also knowing their function and, importantly, their structure–property relationships. This is where the gap is in our present knowledge. Silk proteins have been patented by many research groups and companies and been expressed in bacteria, plants and animals (Arcidiacono *et al.*, 2002; Fahnestock and Bedzyk, 1997; Scheller and Conrad, 2005; Menassa *et al.*, 2004; Lazaris *et al.*, 2002; Karatzas *et al.*, 2005; Huemmerich *et al.*, 2004). But no one, to our knowledge, has succeeded in successfully configuring, i.e. spinning, those proteins into anything resembling the natural fibre, either in its microstructure (which is rather complex) or in its mechanical properties (which are outstanding) (Matsumoto *et al.*, 1996; Shao *et al.*, 2003; Madsen *et al.*, 1999; Marsano *et al.*, 2005; Lazaris *et al.*, 2002; Xie *et al.*, 2006; Liivak *et al.*, 1998; Yao *et al.*, 2002; Zuo *et al.*, 2006; Seidel *et al.*, 2000). The difficulty in creating fibres lies as much in the correct processing conditions, as in the correct protein composition, as we shall discuss in this chapter.

Silks have evolved over hundreds of millions of years and are adaptations to very particular selection pressures (Craig, 1997). Thus, briefly examining the natural function of a particular silk will reveal interesting features about its internal structure which is, after all, what we are trying to copy biotechnologically. Spider silks have excellent mechanical properties, which overall tend to be better than those of any insect silk known (Vollrath and Knight, 2001). For spiders, the use of silk to catch prey has developed in a vigorous 'arms race' with insects during their 400 million years or so of co-evolution (Foelix, 1996). Therefore, both the architecture of the web-trap structure and the building materials, the spider silks, have experienced aeons of very strong optimising selection pressure for toughness (Vollrath and Knight, 2001). On the other hand, insect silks, specifically silk moth silks, are under very different selection pressures, i.e. to build a cocoon in which to pupate from the caterpillar into the moth or butterfly imago (Fedic *et al.*, 2002). For such an 'eggshell', the integration of the fibres into a strong composite is more important than individual fibre toughness itself and, indeed, the structure–function properties of silkworm silks are well matched to this specific task (Zhao *et al.*, 2005).

Micro-morphological studies on the silk production system of the spider reveal that it differs slightly from the Lepidoptera (Kitagawa *et al.*, 2001; Frische *et al.*, 1997; Thiel *et al.*, 1994; Poza *et al.*, 2002). Unlike Lepidopteron insects, which tend to spin a single fibre issuing through the mouth from a merged pair of huge glands, spiders tend to have a whole battery of silks, produced by a multitude of glands, between which they can 'choose' (Foelix, 1996). Every one of these silks has evolved for a particular task, or set of tasks (see Fig. 4.1). This manifests itself in the wide variation of material properties exhibited by different silks, thus serving as excellent indicators of structure–function relationships (Vollrath and Knight, 2001).

Although there are many different types of spider webs and silks, we shall focus our examination on the dragline of a few species (but mainly the golden orb-weaving spiders of the genus *Nephila* sp.), and the comparison of that silk with the cocoon silk of the common (and commercial) silkmoth *Bombyx mori* (hereafter referred to as the silkworm). Other spider silks that are interesting structurally but will have to be ignored here encompass the super-elastomeric (because fully hydrated), micro-windlass capture silks of the ecribellate orb weavers (Vollrath and Edmonds, 1989) or the hackled, nanofilamentous and electrostatic capture silks of the cribellate spiders (Hawthorn and Opell, 2002) as well as the cocoon silks of any spider or the cement silks used to stick down and anchor other silk threads (Foelix, 1996).

The dragline silk produced by the major ampullate glands is not only the principal silk of the 'advanced' spiders, used for both safety and for the guy/anchor threads that suspend the web in mid-air, but it is also the silk that is most easily drawn from the animal under controlled conditions and in enough quantity for comprehensive study (Madsen *et al.*, 1999). Cocoon silk produced by *Bombyx mori* has been artificially selected for over 6000 years, and is produced on an industrial scale equivalent to other high-performance polymers. Therefore, this silk presents an excellent comparative material to that of (mechanically superior) spider silk as it is easy to obtain, available in large quantities and very well characterised.

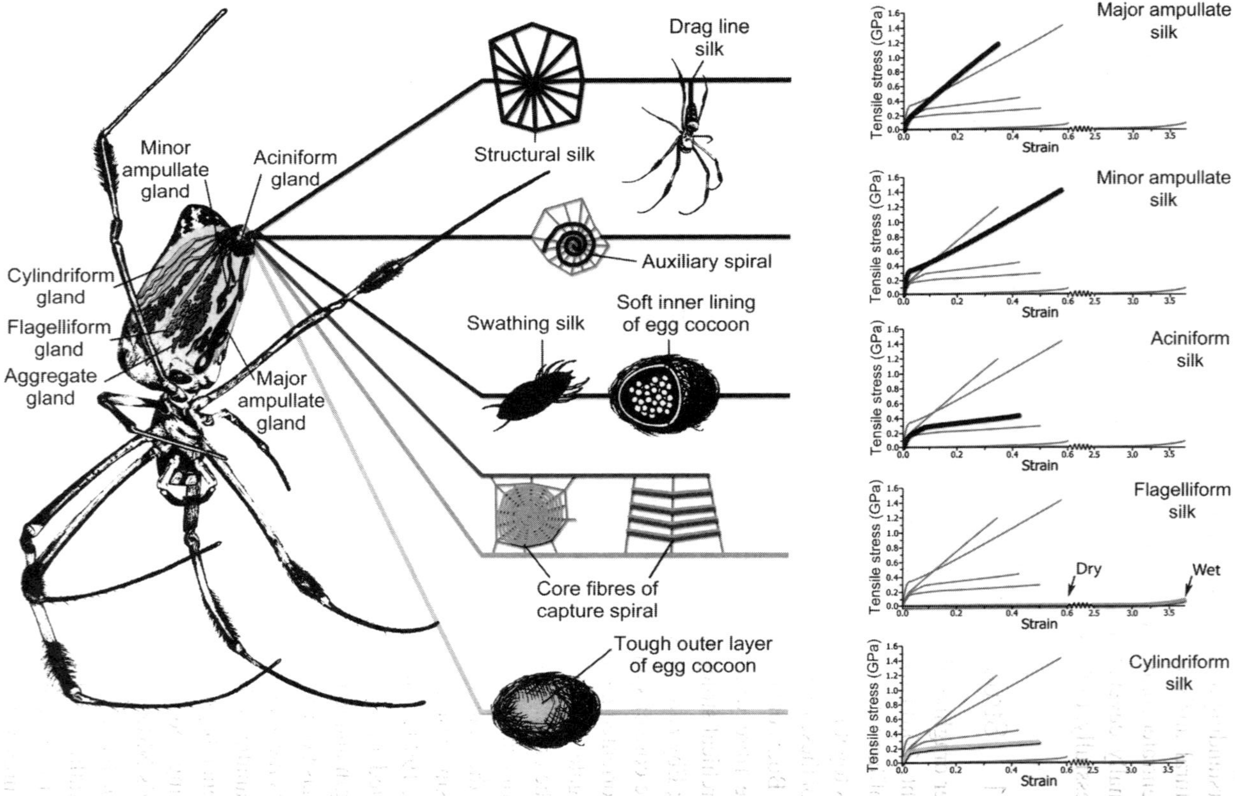

4.1 Glandular origin, function and tensile mechanical properties of silks produced by *Nephila* spiders. Lines are differing shades of grey to match gland of origin and stress–strain plots.

4.2 Unravelling structure–function relationships

Probing structure–function relationships in these materials, requires, initially an understanding of their composition. Silk in particular has leant itself very well to structural analysis as it is one of a few biological materials which is highly concentrated, orientated and designed to perform outside the body, thus techniques originally developed to study man-made fibres can be translated rapidly, and successfully, to the study of silk.

4.2.1 Structure from composition

Spider dragline silk fibres consist of single protein monofilaments (Vollrath and Knight, 2001) whereas the silk of the silkworm consists of a pair of protein monofilaments (brins) coated by a glycoprotein (sericin) used as a matrix to glue fibres together forming the cocoon composite (Poza *et al.*, 2002). Studies on the core of these filaments reveal remarkable similarities between spider and silkworm silks. Both contain nano-fibrillar structures oriented (primarily) along the long axis of the fibre which can form larger structures (microfibrils) thus leading to a hierarchical structure (Frische *et al.*, 1998; Oroudjev *et al.*, 2002; Sapede *et al.*, 2005; Eby *et al.*, 1999).

The core of spider dragline also contains elongated and filled canaliculi 'cavities' oriented parallel to the silk fibre axis and in appearance very similar to the 'elongate vacuolar droplets' of the silk of some moth larva (Akai *et al.*, 1993; Frische *et al.*, 1998). This overall skin-core structure is the result of a rheological pattern originating in the two secreting regions recognised in *Nephila* silk glands with the canaliculi indicating flow inclusions (Knight and Vollrath, 1999c). Stressing a fibre until fracture can lead to cracks forming between canaliculi (Shao *et al.*, 1999a), and it may be that this potentially damaging mechanical energy is diverted on a microscopic level, thus contributing to the exceptional tensile strength and toughness of the fibre.

It has been recently hypothesised that the hierarchical order of structures in silk causes mechanical energy dissipation at not just the microscopic level, as seen in the canaliculi, but by the very nanostructure itself (Porter and Vollrath, 2007). Evidence beginning with the earliest x-ray crystallographic studies in the 1950s (Pauling and Corey, 1951; Lucas *et al.*, 1955; Riekel and Vollrath, 2001) up to today's NMR (Van Beek *et al.*, 2002) and neutron scattering work (Sapede *et al.*, 2005), has revealed silk's molecular architecture to be composed of ordered (crystalline) and disordered (amorphous) domains. This has been refined over the years to a general model of ordered regions, which are composed of β-sheet protein structures with a high degree of internal bonding, and disordered regions, possessing a more heterogeneous distribution of protein structures and bonding (Termonia, 1994; Beek *et al.*, 2002; Porter *et al.*, 2005; Vollrath and Porter, 2006a) (see Fig. 4.2). This combination of, and energetic communication between, ordered

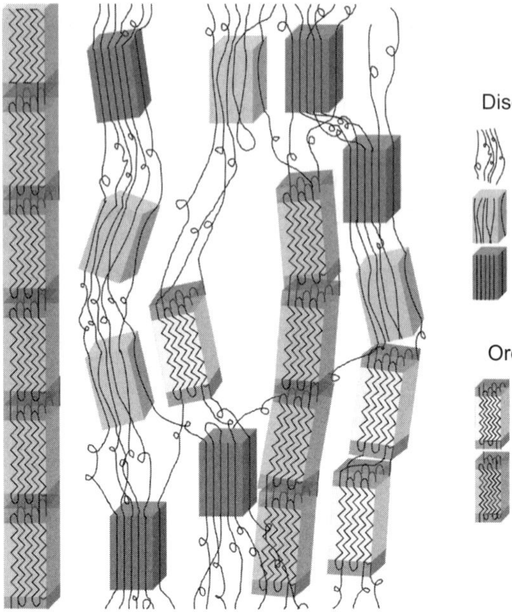

4.2 Diagrammatic representation of the nanostructure of dragline silk. Molecular chains are oriented parallel to the long axis of the fibre and whilst only displayed in two dimensions the network extends into the third. Disordered domains dissipate mechanical energy and are composed of amorphous chains, oriented amorphous chains disrupted by weak polar solvents (such as alcohols), and chains disrupted only by strong polar solvents (such as water). Ordered domains can only be disrupted by strong chaotropic agents and are responsible for the strength of the fibre. These domains consist of β-sheet crystallites whose orientation is either poorly or well defined. Chains of crystallites arrange together to form nanofibrils (far left).

and disordered structures on the nano-scale, enables the dissipation of mechanical energy and reduces local stress concentrations which might precipitate fracture, making silk tough even on the macro-scale (Porter *et al.*, 2005; Porter and Vollrath, 2007).

4.2.2 Structure from processing conditions

Information from static structural studies of silk can be enhanced through the act of controlled perturbation. By applying physical and chemical forces to silk, we can assess how structure *determines* function. For example, the stress–strain characteristics of *Nephila* dragline silk change significantly with processing conditions. Reeling speed (the rate at which silk is experimentally drawn from the spinneret onto a bobbin) as well as body temperature both affect fibre diameter as well as all conceivable mechanical properties (Vollrath *et al.*, 2001), something also observed in silkworm silk (Shao and Vollrath, 2002). Online micro x-ray diffraction (Riekel *et al.*, 2000; Riekel and Vollrath, 2001) or Raman spectroscopy (Young *et al.*, 1998), of single fibres as they are spun, shows corresponding changes in structural parameters such as the degree and orientation of the highly ordered, crystalline domains.

A wide range of solvents are able to modify the material properties of many silks (Shao and Vollrath, 1999; Shao and Vollrath, 1997; Perez-Rigueiro *et al.*, 2001; Perez-Rigueiro *et al.*, 2000). A solvent of particular use in assessing structure–function relationships in dragline silk is water. Dragline silk, unlike other spider silks, undergoes supercontraction when wetted, shrinking to up to half its original length and double its diameter (Work and Morosoff, 1982). This interesting feature has been correlated with a recoverable disruption to partially oriented structures by water in the amorphous domains of the silk (Grubb and Ji, 1999; Simmons *et al.*, 1996; Fornes *et al.*, 1983; Eles and Michal, 2004). The degree to which a dragline silk contracts depends, in part, on the processing conditions, with silks spun at higher rates contracting more due to a higher degree of partially oriented amorphous region for the water to plasticize (Liu *et al.*, 2005). Thus, a fibre's capacity to shrink has therefore been proposed as an indicator of molecular chain orientation; as once supercontracted it can be considered to be in a reference state only slightly affected by processing conditions (Perez-Rigueiro *et al.*, 2003; Liu *et al.*, 2005). Liu *et al.* have argued that supercontraction is actually an evolutionary constraint upon dragline silk, if high strength and toughness are to be achieved then shrinkage when wet has to be accepted (Liu *et al.*, 2005). This is an important factor to note when considering dragline-based biomaterials for use in wet environments.

Previous attempts to link amino acid sequences or protein compositions to mechanical properties have been inconclusive (Zax *et al.*, 2004; Brooks *et al.*, 2005; Brooks *et al.*, 2007; Swanson *et al.*, 2006). However, supercontraction has linked, for the first time, the role of an individual amino acid, proline, to dragline silks' mechanical properties (Liu *et al.*, 2007). By comparing the supercontraction properties of dragline silks differing in amino acid composition, but reeled to have similar breaking strains, Liu *et al.* demonstrated that, whilst processing conditions can alter the capacity of a silk to shrink within a certain range, it is proline content that determines the range in which the capacity to shrink can be adjusted.

It would seem that solvents with different polarities affect different regions of spider silk's composite microstructure (Vollrath *et al.*, 1996), probably by modifying the conformation of the different molecular chains. A comparison of Raman spectra shows clear pre-solvent and solvent-induced differences in a wide range of material properties in the major ampullate dragline silks in four spiders representing different families *Araneus diadematus* (Araneid orb-weavers), *Nephila edulis* (Tetragnathid orb-weavers), *Latrodectus mactans* (Theridiid tangle web-weavers) and *Euprosthenops* sp. (Pisaurid nursery web-weavers) allowing further assignation of conformationally sensitive regions (see Fig 4.2) (Shao *et al.*, 1999b).

Although silkworm silk does not supercontract, water does have a plasticising effect. In order to unravel silk from the cocoon, the sericin matrix embedding the fibres must be removed by 'degumming'. The most common methods of degumming are made possible because of the high content of serine and other hydrophilic amino acids in the sericin proteins (Komatsu, 1975; Gamo *et al.*, 1977) making them soluble in hot alkaline water. Complete removal of sericin has proven to represent a technical trade-off, as prolonged degumming degrades silk's mechanical properties (Perez-Rigueiro *et al.*, 2002), but the slightest presence of sericin can cause problems for the biomedical application of silk-based materials (Panilaitis *et al.*, 2003). Once degummed, immersing silk in water reduces its strength, whilst dehydrating agents, such as acetone and alcohols, increase its stiffness (Perez-Rigueiro *et al.*, 2000) and boiling solutions of salts shrink and can damage the fibre (Tsukada *et al.*, 1994).

The ultimate perturbation of the molecular structure of silk occurs when a solvent interacts with the ordered domains (β-sheet rich), disrupting the amide–amide hydrogen bonds, resulting in a partial/complete breakdown of the fibre. This can only be achieved using highly chaotropic agents such as concentrated solutions of lithium salts (Yamada *et al.*, 2001; Sponner *et al.*, 2005) or halogenated alcohols (Seidel *et al.*, 1998; Fahnestock, 1994). This process is popularly known as regeneration, or perhaps more correctly reconstitution, and is the typical approach to produce a spinning feedstock for artificial silk fibre production.

4.2.3 Structure from modelling

Studies such as those outlined above, aiming to integrate mechanical data with structural information, are important in order to prize apart structure–function relationships. The real value of such studies will become apparent when we are able to integrate information on protein and peptide sequences into this data set. As previously mentioned, we already know that the proline content of a silk correlates well with both the overall silk mechanics as well as supercontraction (Liu *et al.*, 2007), a relationship originally hypothesised in our models (Vollrath and Porter, 2006b). Furthermore, the roles of other silk amino acids (such as glycine) are beginning to be elucidated (Dicko *et al.*, 2007), permitting us to extend our hypotheses to the impact on mechanical behaviour of the different compositions of silk proteins (Vollrath and Porter, 2006b). Clearly, for such insights, modelling is a key tool (Termonia, 1994; Van Nimmen *et al.*, 2005), and recent models are surprisingly good at leading the way towards a comprehensive understanding of silk structure–function relationships (Du *et al.*, 2006; Vollrath and Porter, 2006a; Vollrath and Porter, 2006b; Porter *et al.*, 2005) and even down to the core of protein instability and denaturing (Porter and Vollrath, 2008).

4.3 Spider and worm spinning *in vivo*

So far we have demonstrated that the processing conditions are as important as the material itself, to the extent that even silks with different amino acid compositions can be tuned to have almost identical material properties (Liu *et al.*, 2007). Therefore, successful (i.e. controlled) biomimetic production of these materials will only be possible by understanding not just the material itself but also how the fibres are formed naturally.

Silk is produced and processed by specialist glands in the spider and silkworm in a remarkably similar manner. Silk proteins are released into solution by epithelial cells and stored, potentially for extended periods of time, in a part of the gland serving as a reservoir, known as the ampulla (spiders) or middle division (silkworm) (Knight and Vollrath, 1999a; Akai, 1986). The situation in silkworms is complicated further by the fact that the silk is co-extruded with multiple sericin proteins, produced by a part of the gland further downstream (Gamo *et al.*, 1977). Upon spinning, this protein solution travels down an elongated tapering tubular duct, during which the silk undergoes active chemical and mechanical modification, to convert it from stored gel to final solid fibre as it exits through the spinneret (Vollrath and Knight, 2001; Asakura *et al.*, 2007).

Silk glands can be divided in several zones, each thought responsible for the production of a different combination of silk proteins. In spider dragline silk, this is usually a combination of spidroins (MaSp1 and MaSp2) (Hinman and Lewis, 1992) and, in the silkworm, fibroins (H-fibroin and L-fibroin) and P25 (Shimura *et al.*, 1976; Gamo *et al.*, 1977). Both spidroins and fibroins show a degree of evolutionary convergence in their peptide composition, both possessing high amounts of glycine and alanine in repetitive motifs necessary to propagate the β-sheet structures seen in the fibre (Gatesy *et al.*, 2001; Bini *et al.*, 2004).

Birefringence patterns and the shape changes in the silk droplet inclusions (canaliculi) provided the initial evidence that elongational flow and shear are important in providing the energy for this transition from stored gel to final solid fibre (Knight *et al.*, 2000; Magoshi *et al.*, 1985; Akai, 1986; Akai *et al.*, 1987). There is evidence that this flow is non-Newtonian, resembling nematic liquid-crystalline flow (Knight and Vollrath, 1999a; Knight and Vollrath, 1999b; Asakura *et al.*, 2007) if perhaps not being liquid crystal *sensu strictu* (Holland *et al.*, 2006).

The duct of the silk gland consists of a hollow fibre dialysis membrane that is able to 'handle' the rapid removal of water as well as to precisely control ionic and pH gradients (specifically acidification) during spinning (Foo *et al.*, 2006). This serves to reduce the amount of energy required to undergo a stress-induced phase transition and helps propagate and stabilise the different conformations and arrangements of silk proteins in order to bring about the development of a hierarchical structure (Zong *et al.*, 2004; Magoshi *et al.*, 1997; Dicko *et al.*, 2004a; Dicko *et al.*, 2004b; Dicko *et al.*, 2004c; Vollrath *et al.*, 1998; Knight and Vollrath, 2001; Meyer and Jeannerat, 1939).

The silk gland duct in a spider terminates in a structure [originally termed 'valve' (Wilson, 1962; Wilson, 1969)] because it was thought to work much like the 'press' in the silkworm, i.e. to squeeze the silk and provide shear stress. Whilst this is no longer thought to be used as a 'press' in the spider, it is now seen as a 'ratchet', used to advance broken threads internally (Knight *et al.*, 2000; Knight and Vollrath, 1999b). The 'press' in the silkworm is thought to allow for slight modifications to the degree of post draw applied to the silk because of its relatively narrow range of spinning rates (Magoshi *et al.*, 1985; Iizuka, 1966), whereas the spider can achieve this by altering the rate at which it draws silk with either its legs (if used) or movement of the body (Vollrath *et al.*, 2001). Finally, the thread is stripped of its bathing and coating liquids (save for the sericin in the silkworm) before exiting at a tight spinneret.

Therefore, the process of spinning occurs when silk proteins are pulled through the spinning duct, resulting in a stress-induced phase transition and hierarchical structure development. Other factors, such as the chemical changes to the silk's environment, dehydration and the 'ratchet' or 'press', serve to facilitate this transition or tune the properties of the final fibre. By characterising exactly how these materials respond to shear deformation (mechanical energy input), we can determine how this transition from stored gel to final solid fibre occurs. If we want to spin these materials biomimetically, then an understanding of the stress-induced phase transition will be THE key to controlled processing in order to optimise material properties. By employing rheology, the study of flow and deformation of matter, we are able to examine the spinning forces applied to the feedstock in the duct to samples *in vitro*, providing us with a window into the silk production process.

The first rheological investigations into silk began in the 1960s, but technical constraints meant that the experiments required large sample volumes. This spawned two approaches; either dilution of the native spinning feedstock or using a reconstituted silk, each one moving further away from the natural system (Iizuka, 1966; Ochi *et al.*, 2002; Hossain *et al.*, 2003; Chen *et al.*, 2002). The introduction of more sensitive machines made it possible to start using samples at *in vivo* concentrations (20–30% dry weight) (Terry *et al.*, 2004; Kojic *et al.*, 2006; Holland *et al.*, 2006). However, the ability to test a silk feedstock 'straight from the gland' did bring its own set of challenges, specifically concerning repeatability. When previously working with dilute or reconstituted samples, shear history was inconsequential. However, as silk has evolved to be stored at the precipice of a stress-induced phase transition (to make spinning as fast and energetically efficient as possible) any mechanical energy accidentally introduced during an experiment, by mishandling the samples, could cause up to two orders of magnitude variation between experiments (Holland *et al.*, 2006; Terry *et al.*, 2004).

Rheological studies on native concentrations of silkworm feedstock have revealed that the material behaves like a non-Newtonian high molecular weight polymer acting as a weak gel (Terry *et al.*, 2004, Holland *et al.*, 2006). Unspun silk

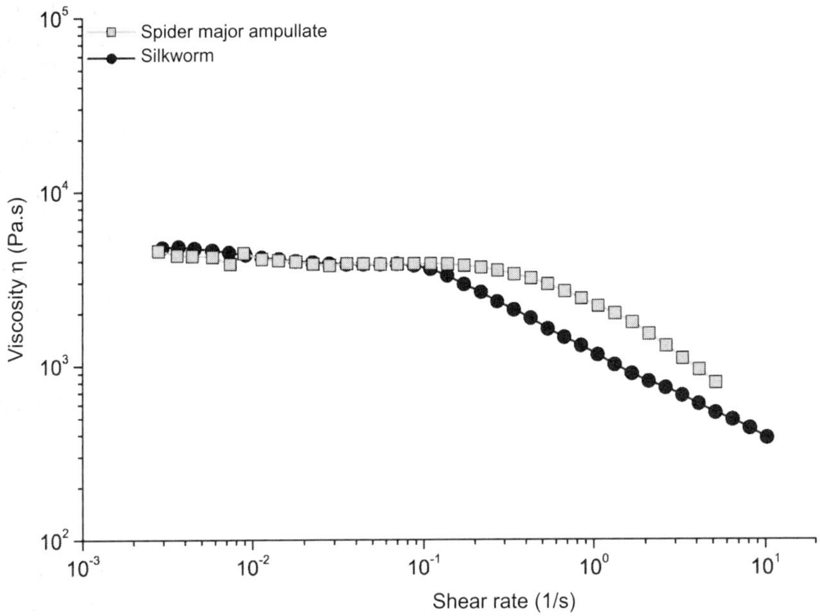

4.3 Viscosity shear rate profiles of spider (squares) and silkworm (circles) native spinning feedstocks. Note the similarity between the profiles in their response to increasing shear.

behaves as a liquid over long timescales (ideal for storage) and a solid over short timescales (suitable for energy absorption for phase transition). In addition, exposing a silk feedstock to a drop in pH results in a rapid gelation of the material, which is synonymous with the acidification observed in the duct (Terry *et al.*, 2004; Vollrath *et al.*, 1998; Magoshi *et al.*, 1994). Later work directly compared spider major ampullate and silkworm silk feedstocks (Holland *et al.*, 2006; Kojic *et al.*, 2006). The findings revealed that these two materials, despite an evolutionary separation of hundreds of millions of years and being made from completely different proteins and making fibres with very different mechanical properties, both share almost identical shear rheologies (see Fig. 4.3) (Holland *et al.*, 2006). This implies that in order to produce a high-performance fibre naturally, these materials have to flow and absorb energy in a similar manner, which might very well be a key evolutionary constraint for silk spinning. In addition, these materials not only behave rheologically like one another, they also behave like molten polymers under flow. This allows us, for the first time, to use tools and techniques originally developed for the polymer industry to study silks, but also serves as a welcome affirmation that we may, one day, be able to adapt traditional polymer spinning technologies to artificially produce silk.

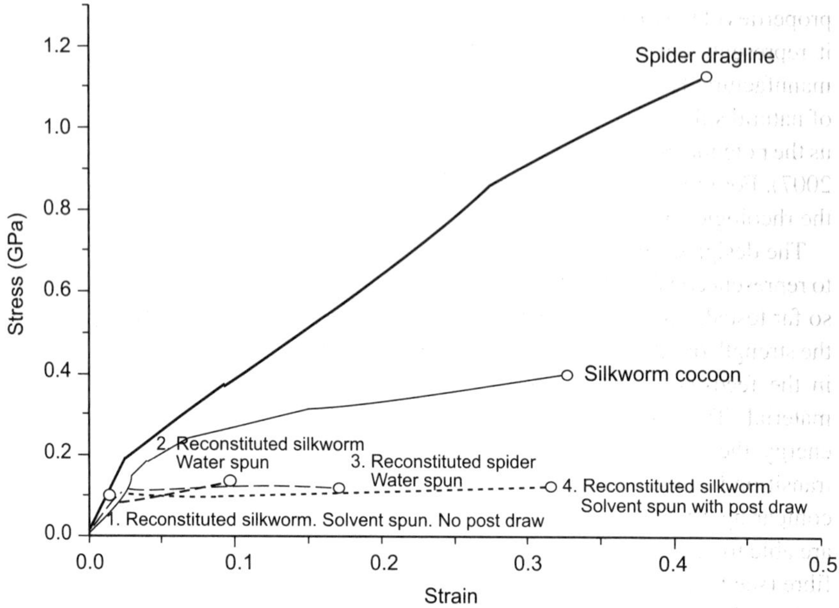

4.4 The tensile properties of artificial silks compared to native spider and silkworm. Labels describing the natural progenitor and spinning conditions are next to the reconstituted stress–strain curves (Madsen *et al.*, 1999; Xie *et al.*, 2006; Shao *et al.*, 2003; Marsano *et al.*, 2005).

4.4 Spinning *in vitro*

Biotechnologically, the spinning pathway of the major ampullate silk in any spider is a highly advanced fibre production system. It is able to control the energetically efficient production of fibres, with a wide variation of mechanical properties, using the same spinning feedstock, by just altering the spinning conditions. These properties, perhaps even above the high performance characteristics, make silk a highly desirable material in order to spin biomimetically.

Many attempts have been made to produce artificial silk, e.g. Matsumoto *et al.*, 1996; Shao *et al.*, 2003; Madsen *et al.*, 1999; Marsano *et al.*, 2005; Lazaris *et al.*, 2002; Xie *et al.*, 2006; Liivak *et al.*, 1998; Yao *et al.*, 2002; Zuo *et al.*, 2006; Seidel *et al.*, 2000, although none can be considered truly biomimetic (i.e. spun in a natural way). Typically such artificial silk fibres are spun from silk that was reconstituted 'back' into its unspun state and then re-processed in a variety of different ways, ranging from classical alcohol baths (Matsumoto *et al.*, 1996) to electrospinning (Zarkoob *et al.*, 2004). However, to date none of these reconstituted fibres have been able to match the mechanical properties of their natural progenitors (see Fig. 4.4).

Determining the quality of artificial silks by testing only the mechanical properties of the final fibre can only provide half the story, as, like the natural fibre, it represents both the quality of the spinning feedstock and the conditions of manufacture. However, from previous work, it has been shown that the rheology of natural silks converges upon a narrow range of rheological parameters, giving us the potential to turn *evolutionary constraints* into *design criteria* (Holland *et al.*, 2007). For biomimetic spinning it will be necessary, so we believe, to first match the rheologies of reconstituted and native feedstocks.

The design criteria are rheological parameters (spinnability indicators) chosen to represent consistent, yet relevant, features seen across the native silk feedstocks so far tested. The first spinnability indicator is *zero shear viscosity*, representing the strength of intermolecular associations (internal friction) between silk proteins in the feedstock. This describes how applied shear energy flows through the material. The second indicator is *plateau modulus,* which reveals how much energy the silk proteins are able to absorb, which is required for a complete transition from stored gel to final solid fibre (Knight *et al.*, 2000). Therefore, by comparing artificial silk feedstocks against these rheological design criteria, we are able to determine the *potential* of a silk to be spun, like a silk, into a silk-like fibre (see Fig. 4.5).

As evidenced from Fig. 4.5 it is clear that native silkworm and artificial reconstituted silkworms spinning feedstocks are significantly different from one another. Clearly, the act of reconstitution seriously degrades not only silk's ability to store energy, required for phase transition, but also degrades or destroys vital associations between silk molecules, required for a hierarchical structure. These flow properties (indicative of *function*) can be related to *structure*. Indeed, the research by Holland and collaborators confirms previous findings about the damaging effects of reconstitution on silk which is seen structurally by a change in conformation and size of the proteins (Yamada *et al.*, 2001; Iridag and Kazanci, 2006; Zuo *et al.*, 2006; Asakura *et al.*, 1985).

Therefore, it is not surprising that industrial methods, such as alcohol baths, extreme chemical modification and temperature are required in order to 'spin' a reconstituted silk fibre (Matsumoto *et al.*, 1996; Shao *et al.*, 2003; Madsen *et al.*, 1999; Marsano *et al.*, 2005; Lazaris *et al.*, 2002; Xie *et al.*, 2006; Liivak *et al.*, 1998; Yao *et al.*, 2002; Zuo *et al.*, 2006; Seidel *et al.*, 2000). These approaches are the only way to introduce sufficient energy into these materials to bring about fibre formation due to the silk protein's severely degraded nature. The massive gap between these two feedstocks may go some way to explaining why it has not been possible to create a reconstituted 'silk' fibre with the mechanical properties (Matsumoto *et al.*, 1996; Shao *et al.*, 2003; Madsen *et al.*, 1999; Marsano *et al.*, 2005; Lazaris *et al.*, 2002; Xie *et al.*, 2006; Liivak *et al.*, 1998; Yao *et al.*, 2002; Zuo *et al.*, 2006), or structural complexity (Eby, 1995; Putthanarat *et al.*, 2000), of a natural silk, let alone process it in the same way. Clearly, artificial silks are inherently different before they are even spun and it is questionable whether, once reconstituted, these materials can even be called silks.

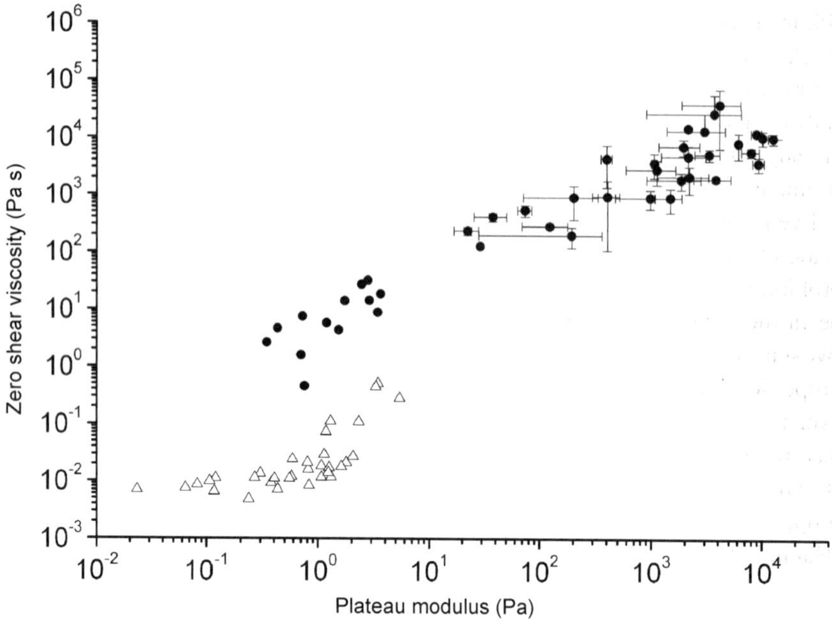

4.5 Spinnability indicators of a similar range of concentrations for native silkworm spinning feedstock (circles), compared with those of artificial, reconstituted silkworm silk feedstock (triangles). Zero shear viscosity indicates the degree of intermolecular associations between silk molecules (for structure development) whilst plateau modulus indicates the amount of energy the molecules can absorb (necessary for phase transition). Error bars for native silkworm spinning feedstock samples >1 Pa, represent standard error based on 3 repeats from the same silkworm gland contents, lower concentration native and reconstituted silkworm spinning silk points represent individual tests due to the increased amount of sample required for accurate characterisation.

4.5 Future trends and applications

As far as we know, none of these enterprises devoted to the commercial production of silk-based materials has managed to make a feedstock that has been spun (or that can be spun) into fibres with properties comparable with those of the role-model, spider dragline silk. We argue that the most logical way forward is to improve the quality of the starting material to match the rheological *design criteria* already provided to us by Nature herself. Future efforts will need to be focused towards a better understanding of what is happening to silk proteins during reconstitution (Holland *et al.*, 2007, Porter and Vollrath, 2008). Through the development of milder, less damaging reconstitution techniques, more benign spinning techniques can be employed and higher performance fibres created (Z. Shao *personal communication*). Another approach is to harness genetic engineering to produce silk feedstock *ab initio*, thus side-stepping the problems of reconstituting a silk 'back' into its unspun form (Vendrely and Scheibel, 2007). Whilst not without their own set of technical difficulties, current attempts at producing such genetically engineered feedstocks are promising, with recombinant spider silk hydrogels already displaying some of the rheological characteristics of natural silk (Rammensee *et al.*, 2006).

Even when truly biomimetic silk production becomes a reality, it is likely that, at least for the foreseeable future, the costs of such fibres will be high and probably prohibitively high for the mass textile market. The initial market will most likely be in the medical field, both for single threads (suture) and specialist textiles (woven and non-woven) which will make use of silks, excellent mechanical properties, biocompatibility and its ability to be chemically functionalised. However, the potential uses of silk are not strictly limited to a fibrous role. Reconstitution has the unique advantage that it allows silk to be reprocessed into almost any structure, from films to foams, whilst still possessing very good mechanical properties, finding uses in areas of medicine from wound dressings to orthopaedics (Hakimi *et al.*, 2007; Vepari and Kaplan, 2007).

4.6 Conclusions

Spider silks, like silkworm silks, are semicrystalline biopolymers with excellent mechanical properties. Independently, spiders and insects have evolved materials that behave virtually identically rheologically but make fibres with very different mechanical characteristics. This is due partly to differences in chemical composition and partly to differences in the details of the processing conditions. Spider silks, because of their selection in a ferocious arms race with insects, have evolved into materials having a wide range of mechanical properties. This makes them very interesting as models for the design and production of modern artificial synthetic silks. The spider's production system has strong potential to provide us with excellent fibres that are produced along biological principles, i.e. ecologically compatible, therefore it is likely that, in the present economic climate, efforts will be redoubled to produce commercial silks the spider's way.

4.7 Sources of further information and advice

Further information can be found within the references for this article. The first piece of literature before embarking on a study of this topic must be Craig's comprehensive review of the evolution of arthropod silks (Craig, 1997). A good overview of the biology of the spider and the silkworm is given by Foelix (1996) and Fedic *et al.* (2002). For a general introduction to silk processing in spiders we would suggest that by Vollrath and Knight (2001) and for silkworms that by Asakura *et al.* (2007). For introductory papers on nano-scale toughness see Porter and Vollrath (2007), for supercontraction see Liu *et al.* (2007), for the effects of solvents see Shao and Vollrath (1999), for reconstitution see Shao *et al.* (2003) and Yamada *et al.* (2001), for structure–function modelling see Vollrath and Porter (2006a) and Vollrath and Porter (2006b), and for silk rheology see Holland *et al.* (2006 and 2007) and Ferry (1980). For a review of genetic engineering of spider silks see Vendrely and Scheibel (2007). Of particular importance outside of the field of silks, yet born of its research and with application to all proteins, the authors suggest the work of Porter and Vollrath (2008).

For applications of silkworm silk as a biomedical material the recent review by Vepari and Kaplan (2007) is of note as is that of Hakimi *et al.* (2007), which deals with both spider and silkworm silk.

Websites

Some research groups working within the field of silk

Prof. Fritz Vollrath, Department of Zoology, Oxford University, Oxford, UK. http://www.oxfordsilkgroup.com
Prof. Zhengzhong Shao, Department of Macromolecular Science, Fudan University, Shanghai, China. http://www.polymer.fudan.edu.cn/research/shaozz/English%20version/index.html
Prof. David Kaplan, Department of Biomedical Engineering, Tufts University. Medford, MA, USA. http://ase.tufts.edu/biomedical/faculty-staff/kaplan.asp
Dr Thomas Scheibel, Technische Universität Munich, Munich, Germany. http://www.fiberlab.de/

Companies producing silk-based biomaterials

Oxford Biomaterials. http://www.oxfordbiomaterials.com
Nexia Biotechnologies Ltd. http://www.nexiabiotech.com

4.8 Acknowledgements

We thank the European Commission, the Royal Society of London, Jesus College Oxford and the US-AFOSR for funding.

4.9 References

Akai, H. (1986) Scanning electron microscopic observations of fibroin secretion into the lumen of the silk gland in Bombyx mori. *Journal of Sericulture Science of Japan*, **55**, 163–167.

Akai, H., Imai, T. and Tsubouchi, K. (1987) Fine-structural changes of liquid silk in the silk gland during spinning stage of Bombyx larvae. *Journal of Sericulture Science of Japan*, **56**, 131–137.

Akai, H., Nagashima, T. and Aoyagi, S. (1993) Ultrastructure of posterior silk gland-cells and liquid silk in Indian Tasar silkworm, Antheraea-Mylitta Drury (Lepidoptera, Saturniidae). *International Journal of Insect Morphology and Embryology*, **22**, 497–506.

Arcidiacono, S., Mello, C. M., Butler, M., Welsh, E., Soares, J. W., Allen, A., Ziegler, D., Laue, T. and Chase, S. (2002) Aqueous processing and fiber spinning of recombinant spider silks. *Macromolecules*, **35**, 1262–1266.

Asakura, T., Kuzuhara, A., Tabeta, R. and Saito, H. (1985) Conformational characterization of Bombyx mori silk fibroin in the solid state by high-frequency carbon-13 cross polarization-magic angle spinning NMR, x-ray diffraction, and infrared spectroscopy. *Macromolecules*, **18**, 1841–1845.

Asakura, T., Umemura, K., Nakazawa, Y., Hirose, H., Higham, J. and Knight, D. (2007) Some observations on the structure and function of the spinning apparatus in the silkworm Bombyx mori. *Biomacromolecules*, **8**, 175–181.

Beek, V. J. D., Hess, H., Vollrath, F. and Meier, B. H. (2002) The molecular structure of spider dragline silk: Folding and orientation of the protein backbone. *Proceedings of the National Academy of Sciences of the United States of America*, **99**, 10266–10271.

Bini, E., Knight, D. P. and Kaplan, D. L. (2004) Mapping domain structures in silks from insects and spiders related to protein assembly. *Journal of Molecular Biology*, **335**, 27–40.

Brooks, A. E., Steinkraus, H. B., Nelson, S. R. and Lewis, R. V. (2005) An investigation of the divergence of major ampullate silk fibers from Nephila clavipes and Argiope aurantia. *Biomacromolecules*, **6**, 3095–3099.

Brooks, A. E., Steinkraus, H. B., Nelson, S. R. and Lewis, R. V. (2007) An investigation of the divergence of major ampullate silk fibers from Nephila clavipes and Argiope aurantia. Erratum. *Biomacromolecules*, **8**, 1358.

Chen, X., Knight, D. P. and Vollrath, F. (2002) Rheological characterization of Nephila spidroin solution. *Biomacromolecules*, **3**, 644–648.

Craig, C. (1997) Evolution of arthropod silks. *Annual Reviews of Entomology*, **42**, 231–267.

Dicko, C., Kenney, J. M., Knight, D. and Vollrath, F. (2004a) Transition to a beta-sheet-rich structure in spidroin in vitro: the effects of pH and cations. *Biochemistry*, **43**, 14080–14087.

Dicko, C., Knight, D., Kenney, J. M. and Vollrath, F. (2004b) Structural conformation of spidroin in solution: a synchrotron radiation circular dichroism study. *Biomacromolecules*, **5**, 758–767.

Dicko, C., Porter, D., Bond, J. E., Kenney, J. and Vollrath, F. (2007) Structural disorder in silk proteins reveal the emergence of silk elastomericity. *Biomacromolecules*, **9**, 216–221.

Dicko, C., Vollrath, F. and Kenney, J. M. (2004c) Spider silk protein refolding is controlled by changing pH. *Biomacromolecules*, **5**, 704–710.

Du, N., Xiang, Y. L., Narayanan, J., Li, L., Lim, M. L. M. and Li, D. (2006) Design of superior spider silk: From nanostructure to mechanical properties. *Biophysical Journal*, **91**, 4528–4535.

Eby, R. K. (1995) Structure and properties of spider dragline. *American Crystallography Association*, **2**(23), 68.

Eby, R. K., Putthanarat, S. and Miller, L. D. (1999) The nanofibrillar morphology of silks. *Abstracts of Papers of the American Chemical Society*, **217**, 069-BTEC.

Eles, P. T. and Michal, C. A. (2004) Strain dependent local phase transitions observed during controlled supercontraction reveal mechanisms in spider silk. *Macromolecules*, **37**, 1342–1345.

Fahnestock, S. (1994) Novel recombinantly formed spider silk analogs. USA, Du Pont.

Fahnestock, S. R. and Bedzyk, L. A. (1997) Production of synthetic spider dragline silk protein in Pichia pastoris. *Applied Microbiology and Biotechnology*, **47**, 33–39.

Fedic, R., Zurovec, M. and Sehnal, F. (2002) The silk of lepidoptera. *Journal of Insect Biotechnology and Sericology*, **71**, 1–15.

Ferry, J. D. (1980) *Viscoelastic properties of polymers*, New York, Wiley VCH.

Foelix, R. F. (1996) *Biology of Spiders (2nd edition)*, Oxford University Press.

Foo, C. W. P., Bini, E., Hensman, J., Knight, D. P., Lewis, R. V. and Kaplan, D. L. (2006) Role of pH and charge on silk protein assembly in insects and spiders. *Applied Physics A: Materials Science and Processing*, **82**, 223–233.

Fornes, R. E., Work, R. W. and Morosoff, N. (1983) Molecular-orientation of spider silks in the natural and supercontracted states. *Journal of Polymer Science Part B-Polymer Physics*, **21**, 1163–1172.

Frische, S., Maunsbach, A. and Vollrath, F. (1998) Elongate cavities and skin-core structure in Nephila spider silk observed by electron microscopy. *Journal of Microscopy*, **189**, 64–70.

Frische, S., Maunsbach, A. and Vollrath, F. (1997) Microstructure of two spider silks. *Journal of Microscopy*, **189**, 64–70.

Gamo, T., Inokuchi, T. and Laufer, H. (1977) Polypeptides of fibroin and sericin secreted from the different sections of the silk gland in Bombyx mori. *Insect Biochemistry*, **7**, 285–295.

Gatesy, J., Hayashi, C., Motriuk, D., Woods, J. and Lewis, R. (2001) Extreme diversity, conservation, and convergence of spider silk fibroin sequences. *Science*, **291**, 2603–2605.

Grubb, D. T. and Ji, G. D. (1999) Molecular chain orientation in supercontracted and re-extended spider silk. *International Journal of Biological Macromolecules*, **24**, 203–210.

Hakimi, O., Knight, D. P., Vollrath, F. and Vadgama, P. (2007) Spider and mulberry silkworm silks as compatible biomaterials. *Composites Part B: Engineering*, **38**, 324–337.

Hawthorn, A. C. and Opell, B. D. (2002) Evolution of adhesive mechanisms in cribellar spider prey capture thread: evidence for van der Waals and hygroscopic forces. *Biological Journal of the Linnean Society*, **77**, 1–8.

Hinman, M. B. and Lewis, R. V. (1992) Isolation of a clone encoding a 2nd dragline silk fibroin – Nephila-clavipes dragline silk is a 2-protein fiber. *Journal of Biological Chemistry*, **267**, 19320–19324.

Holland, C., Terry, A. E., Porter, D. and Vollrath, F. (2007) Natural and unnatural silks. *Polymer*, **48**, 3388–3392.

Holland, C. A., Terry, A. E., Porter, D. and Vollrath, F. (2006) Comparing the rheology of native spider and silkworm spinning dope. *Nature Materials*, **5**, 870–874.

Hossain, K. S., Ohyama, E., Ochi, A., Magoshi, J. and Nemoto, N. (2003) Dilute-solution properties of regenerated silk fibroin. *Journal of Physical Chemistry B*, **107**, 8066–8073.

Huemmerich, D., Scheibel, T., Vollrath, F., Cohen, S., Gat, U. and Ittah, S. (2004) Novel assembly properties of recombinant spider dragline silk proteins. *Current Biology*, **14**, 2070–4.

Iizuka, E. (1966) Mechanism of fiber formation by the silkworm Bombyx Mori. *Biorheology*, **3**, 141–152.

Iridag, Y. and Kazanci, M. (2006) Preparation and characterization of *Bombyx mori* silk fibroin and wool keratin. *Journal of Applied Polymer Science*, **100**, 4260–4264.

Karatzas, C. N., Chretien, N., Duguay, F., Bellemare, A., Zhou, Y. F., Rodenhiser, A., Islam, A. S., Turcotte, C., Huang, Y. and Lazaris, A. (2005) High-toughness spider silk fibers spun from soluble recombinant silk produced in mammalian cells. In Steinbuchel, A. and Doi, Y. (Eds.) *Biotechnology of biopolymers; from synthesis to patents Volume 2.* Wiley-VCH.

Kitagawa, M., Yasutomi, M. and Furukawa, A. (2001) Microstructure of spider dragline. *Zairyo*, **50**, 1213–1217.

Knight, D. and Vollrath, F. (1999a) Hexagonal columnar liquid crystal in the cells secreting spider silk. *Tissue and Cell*, **31**, 617–620.

Knight, D. P., Knight, M. M. and Vollrath, F. (2000) Beta transition and stress-induced phase separation in the spinning of spider dragline silk. *International Journal of Biological Macromolecules*, **27**, 205–210.

Knight, D. P. and Vollrath, F. (1999b) Liquid crystals and flow elongation in a spider's silk production line. *Proceedings of the Royal Society of London Series B – Biological Sciences*, **266**, 519–523.

Knight, D. P. and Vollrath, F. (1999c) Liquid crystals and flow elongation in a spider's silk production line. *Proceedings of the Royal Society – Biological Sciences (Series B)*, **266**, 519–523.

Knight, D. P. and Vollrath, F. (2001) Changes in element composition along the spinning duct in a Nephila spider. *Naturwissenschaften*, **88**, 179–182.

Kojic, N., Bico, J., Clasen, C. and McKinley, G. H. (2006) Ex vivo rheology of spider silk. *Journal of Experiemental Biology*, **209**, 4355–4362.

Komatsu, K. (1975) Studies on dissolution behaviors and structural characteristics of silk sericin. *Bull. Sericul. Exp. Sta.*, **26**, 135–256.

Lazaris, A., Arcidiacono, S., Huang, Y., Zhou, J. F., Duguay, F., Chretien, N., Welsh, E. A., Soares, J. W. and Karatzas, C. N. (2002) Spider silk fibers spun from soluble recombinant silk produced in mammalian cells. *Science*, **295**, 472–476.

Liivak, O., Blye, A., Shah, N. and Jelinski, L. W. (1998) A microfabricated wet-spinning apparatus to spin fibers of silk proteins. Structure-property correlations. *Macromolecules*, **31**, 2947–2951.

Liu, Y., Shao, Z. and Vollrath, F. (2005) Relationships between supercontraction and mechanical properties of spider silk. *Nature Materials*, **4**, 901–905.

Liu, Y., Sponner, A., Porter, D. and Vollrath, F. (2008) Proline and processing of spider silks. *Biomacromolecules*, **9**(1), 116–121.

Lucas, F., Shaw, J. and Smith, S. (1955) The chemical constitution of some silk fibroins and its bearing on their physical properties. *Journal of the Textile Institute*, **46**, T440–T452.

Madsen, B., Shao, Z. Z. and Vollrath, F. (1999) Variability in the mechanical properties of spider silks on three levels: interspecific, intraspecific and intraindividual. *International Journal of Biological Macromolecules*, **24**, 301–306.

Magoshi, J., Magoshi, Y., Kato, M., Becker, M. A. and Nakamura, S. (1997) The role of calcium ions in the silk fibroin fiber formation process. *Abstracts of Papers of the American Chemical Society*, **214**, 209–PMSE.

Magoshi, J., Magoshi, Y. and Nakamura, S. (1985) Crystallization, liquid-crystal, and fiber formation of silk fibroin. *Applied Polymer Symposia*, 187–204.

Magoshi, J., Magoshi, Y. and Nakamura, S. (1994) Mechanism of fiber formation of silkworm. In D. L. Kaplan *et al.* (Eds), *Silk polymers: materials science and biotechnology*, ACS Books, Washington.

Marsano, E., Corsini, P., Arosio, C., Boschi, A., Mormino, M. and Freddi, G. (2005) Wet spinning of Bombyx mori silk fibroin dissolved in N-methyl morpholine N-oxide and properties of regenerated fibres. *International Journal of Biological Macromolecules*, **37**, 179–188.

Matsumoto, K., Uejima, H., Iwasaki, T., Sano, Y. and Sumino, H. (1996) Studies on regenerated protein fibers. 3. Production of regenerated silk fibroin fiber by the self-dialyzing wet spinning method. *Journal of Applied Polymer Science*, **60**, 503–511.

Menassa, R., Zhu, H., Karatzas, C. N., Lazaris, A., Richman, A. and Brandle, J. (2004) Spider dragline silk proteins in transgenic tobacco leaves: accumulation and field production. *Plant Biotechnology Journal*, **2**, 431–438.

Meyer, K. H. and Jeannerat, J. (1939) Les propriétés des polymères en solution XI. Sur la formation du fil de soie à partir du continu liquide de la glande. *Helvetica Chimica Acta*, **22**, 22–30.

Ochi, A., Hossain, K. S., Magoshi, J. and Nemoto, N. (2002) Rheology and dynamic light scattering of silk fibroin solution extracted from the middle division of Bombyx mori silkworm. *Biomacromolecules*, **3**, 1187–1196.

Oroudjev, E., Soares, J., Arcdiacono, S., Thompson, J. B., Fossey, S. A. and Hansma, H. G. (2002) Segmented nanofibers of spider dragline silk: atomic force microscopy and single-molecule force spectroscopy. *Proceedings of the National Academy of Sciences of the United States of America*, **99**, Suppl. 2, 6460–6465.

Panilaitis, B., Altman, G. H., Chen, J., Jin, H.-J., Karageorgiou, V. and Kaplan, D. L. (2003) Macrophage responses to silk. *Biomaterials*, **24**, 3079–3085.

Pauling, L. and Corey, R. B. (1951) Configurations of polypeptide chains with favored orientations around single bonds: two new pleated sheets. *Proceedings of the National Academy of Sciences*, **37**, 729–740.

Perez-Rigueiro, J., Elices, M. and Guinea, G. V. (2003) Controlled supercontraction tailors the tensile behaviour of spider silk. *Polymer*, **44**, 3733–3736.

Perez-Rigueiro, J., Elices, M., Llorca, J. and Viney, C. (2001) Tensile properties of Attacus atlas silk submerged in liquid media. *Journal of Applied Polymer Science*, **82**, 53–62.

Perez-Rigueiro, J., Elices, M., Llorca, J. and Viney, C. (2002) Effect of degumming on the tensile properties of silkworm (Bombyx mori) silk fiber. *Journal of Applied Polymer Science*, **84**, 1431–1437.

Perez-Rigueiro, J., Viney, C., Llorca, J. and Elices, M. (2000) Mechanical properties of silkworm silk in liquid media. *Polymer*, **41**, 8433–8439.

Porter, D. and Vollrath, F. (2007) Nanoscale toughness of spider silk. *Nanotoday*, **2**, 6.

Porter, D. and Vollrath, F. (2008) The role of kinetics of water and amide bonding in protein stability. *Soft Matter*, **4**, 328–336.

Porter, D., Vollrath, F. and Shao, J. Z. (2005) Predicting the mechanical properties of spider silk as a model nanostructured polymer. *European Physical Journal E*, **16**, 199–206.

Poza, P., Perez-Rigueiro, J., Elices, M. and Llorca, J. (2002) Fractographic analysis of silkworm and spider silk. *Engineering Fracture Mechanics*, **69**, 1035–1048.

Putthanarat, S., Stribeck, N., Fossey, S. A., Eby, R. K. and Adams, W. W. (2000) Investigation of the nanofibrils of silk fibers. *Polymer*, **41**, 7735–7747.

Rammensee, S., Huemmerich, D., Hermanson, K. D., Scheibel, T. and Bausch, A. R. (2006) Rheological characterization of hydrogels formed by recombinantly produced spider silk. *Applied Physics A: Materials Science and Processing*, **82**, 261–264.

Riekel, C., Madsen, B., Knight, D. and Vollrath, F. (2000) X-ray diffraction on spider silk during controlled extrusion under a synchrotron radiation x-ray beam. *Biomacromolecules*, **1**, 622–626.

Riekel, C. and Vollrath, F. (2001) Spider silk fibre extrusion: combined wide- and small-angle x-ray microdiffraction experiments. *International Journal of Biological Macromolecules*, **29**, 203–210.

Sapede, D., Seydel, T., Forsyth, V. T., Koza, M. M., Schweins, R., Vollrath, F. and Riekel, C. (2005) Nanofibrillar structure and molecular mobility in spider dragline silk. *Macromolecules*, **38**, 8447–8453.

Scheller, J. and Conrad, U. (2005) Spider silk proteins from transgenic plants. In Steinbuchel, A. and Doi, Y. (Eds) *Biotechnology of biopolymers; from synthesis to patents Volume 2*. Wiley-VCH.

Seidel, A., Liivak, O., Calve, S., Adaska, J., Ji, G. D., Yang, Z. T., Grubb, D., Zax, D. B. and

Jelinski, L. W. (2000) Regenerated spider silk: Processing, properties, and structure. *Macromolecules*, **33**, 775–780.

Seidel, A., Liivak, O. and Jelinski, L. W. (1998) Artificial spinning of spider silk. *Macromolecules*, **31**, 6733–6736.

Shao, Z., Hu, X. W., Frische, S. and Vollrath, F. (1999a) Heterogeneous morphology of Nephila edulis spider silk and its significance for mechanical properties. *Polymer*, **40**, 4709–4711.

Shao, Z. and Vollrath, F. (1997) Mechanics of spider silk during and after contraction in selected solvents. *10th Int Conference on Deformation, Yield and Fracture of Polymers*, 262–265.

Shao, Z., Vollrath, F., Sirichaisit, J. and Young, R. J. (1999b) Analysis of spider silk in native and supercontracted states using Raman spectroscopy. *Polymer*, **40**, 2493–2500.

Shao, Z. Z. and Vollrath, F. (1999) The effect of solvents on the contraction and mechanical properties of spider silk. *Polymer*, **40**, 1799–1806.

Shao, Z. Z. and Vollrath, F. (2002) Surprising strength of silkworm silk. *Nature*, **418**, 741.

Shao, Z. Z., Vollrath, F., Yang, Y. and Thogersen, H. C. (2003) Structure and behavior of regenerated spider silk. *Macromolecules*, **36**, 1157–1161.

Shimura, K., Kikuchi, A., Ohtomo, K., Katagata, Y. and Hyodo, A. (1976) Studies on silk fibroin of Bombyx mori. I. Fractionation of fibroin prepared from the posterior silk gland. *Journal of Biochemistry (Tokyo)*, **80**, 693–702.

Simmons, A. H., Michal, C. A. and Jelinski, L. W. (1996) Molecular orientation and two-component nature of the crystalline fraction of spider dragline silk. *Science*, **271**, 84–87.

Sponner, A., Schlott, B., Vollrath, F., Unger, E., Grosse, F. and Weisshart, K. (2005) Characterization of the protein components of Nephila clavipes dragline silk. *Biochemistry*, **44**, 4727–4736.

Swanson, B. O., Blackledge, T. A., Summers, A. P. and Hayashi, C. Y. (2006) Spider dragline silk: correlated and mosaic evolution in high-performance biological materials. *Evolution: International Journal of Organic Evolution*, **60**, 2539–51.

Termonia, Y. (1994) Molecular modeling of spider silk elasticity. *Macromolecules*, **27**, 7378–7381.

Terry, A. E., Knight, D. P., Porter, D. and Vollrath, F. (2004) pH induced changes in the rheology of silk fibroin solution from the middle division of Bombyx mori silkworm. *Biomacromolecules*, **5**, 768–772.

Thiel, B. L., Kunkel, D., Guess, K. and Viney, C. (1994) Composite microstructure of spider (Nephila clavipes) dragline. In M. Alper, H. Bayley, D. Kaplan and M. Navia (Eds) *Biomolecular Materials by Design, MRS Symposium.* Pittsburgh, Materials Research Society.

Tsukada, M., Kato, H., Freddi, G., Kasai, N. and Ishikawa, H. (1994) Structural changes and dyeability of silk fibroin fiber following shrinkage in neutral salt solution. *Journal of Applied Polymer Science*, **51**, 619–624.

Van Beek, J. D., Hesst, S., Vollrath, F. and Meier, B. H. (2002) The molecular structure of spider dragline silk: Folding and orientation of the protein backbone. *Proceedings of the National Academy of Sciences of the United States of America*, **99**, 10266–10271.

Van Nimmen, E., Gellynck, K., Gheysens, T., Van Langenhove, L. and Mertens, J. (2005) Modelling of the stress–strain behavior of egg sac silk of the spider Araneus diadematus. *Journal of Arachnology*, **33**, 629–639.

Vendrely, C. and Scheibel, T. (2007) Biotechnological production of spider-silk proteins enables new applications. *Macromolecular Bioscience*, **7**, 401–409.

Vepari, C. and Kaplan, D. L. (2007) Silk as a biomaterial. *Progress in Polymer Science*, **32**, 991–1007.

Vollrath, F. and Edmonds, D. T. (1989) Modulation of the mechanical properties of spider silk by coating with water. *Nature*, **340**, 305–307.

Vollrath, F., Holtet, T., Thogersen, H. C. and Frische, S. (1996) Structural organization of spider silk. *Proceedings of the Royal Society of London Series B – Biological Sciences*, **263**, 147–151.

Vollrath, F. and Knight, D. P. (2001) Liquid crystalline spinning of spider silk. *Nature*, **410**, 541–548.

Vollrath, F., Knight, D. P. and Hu, X. W. (1998) Silk production in a spider involves acid bath treatment. *Proceedings of the Royal Society of London Series B – Biological Sciences*, **265**, 817–820.

Vollrath, F., Madsen, B. and Shao, Z. (2001) The effect of spinning conditions on the mechanics of a spider's dragline silk. *Proceedings of the Royal Society – Biological Sciences (Series B)*, **268**, 2339–2346.

Vollrath, F. and Porter, D. (2006a) Spider silk as a model biomaterial. *Applied Physics A: Materials Science and Processing*, **82**, 205–212.

Vollrath, F. and Porter, D. (2006b) Spider silk as an archetypal protein elastomer. *Soft Matter*, **2**, 377–385.

Wilson, R. S. (1962) The structure of the dragline control valves in the garden spider. *Quarterly Journal of Microscopical Science*, **103**, 549–555.

Wilson, R. S. (1969) Control of dragline spinning in certain spiders. *American Zoologist*, **9**, 103–111.

Work, R. W. and Morosoff, N. (1982) A physicochemical study of the supercontraction of spider major ampullate silk fibers. *Textile Research Journal*, **52**, 349–356.

Xie, F., Zhang, H., Shao, H. and Hu, X. (2006) Effect of shearing on formation of silk fibers from regenerated Bombyx mori silk fibroin aqueous solution. *International Journal of Biological Macromolecules*, **38**, 284–288.

Yamada, H., Nakao, H., Takasu, Y. and Tsubouchi, K. (2001) Preparation of undegraded native molecular fibroin solution from silkworm cocoons. *Materials Science and Engineering C – Biomimetic Materials Sensors and Systems*, **14**, 41–46.

Yao, J., Masuda, H., Zhao, C. and Asakura, T. (2002) Artificial spinning and characterization of silk fiber from bombyx mori silk fibroin in hexafluoroacetone hydrate. *Macromolecules*, **35**, 6–9.

Young, R. J., Sirichaisit, J., Shao, Z. and Vollrath, F. (1998) Analysis of structure/property relationships in spider silk using Raman spectroscopy. *The New Polymers: Science, Technology and Applications*. Brighton, Institute of Physics.

Zarkoob, S., Eby, R. K., Reneker, D. H., Hudson, S. D., Ertley, D. and Adams, W. W. (2004) Structure and morphology of electrospun silk nanofibers. *Polymer*, **45**, 3973–3977.

Zax, D. B., Armanios, D. E., Horak, S., Malowniak, C. and Yang, Z. (2004) Variation of mechanical properties with amino acid content in the silk of Nephila clavipes. *Biomacromolecules*, **5**, 732–8.

Zhao, H.-P., Feng, X.-Q., Yu, S.-W., Cui, W.-Z. and Zou, F.-Z. (2005) Mechanical properties of silkworm cocoons. *Polymer*, **46**, 9192–9201.

Zong, X., Zhou, P., Shao, Z. Z., Chen, S. M., Chen, X., Hu, B. W., Deng, F. and Yao, W. H. (2004) Effect of pH and copper (II) on the conformation transitions of silk fibroin based on EPR, NMR, and Raman spectroscopy. *Biochemistry*, **43**, 11932–11941.

Zuo, B., Dai, L. and Wu, Z. (2006) Analysis of structure and properties of biodegradable regenerated silk fibroin fibers. *Journal of Materials Science*, **41**, 3357–3361.

5

A biomimetic approach to the production of sustainable structural composites using plant fibres

C. SANTULLI
University of Reading, UK

Abstract: The production of sustainable structural composites, using cellulosic fibres extracted from plants as a reinforcement, is discussed. Critical factors, including fibre extraction and treatment, impact properties, compatibility with polymer matrices, and selection of plant fibres are explored. Ideally, all of these aspects need to be integrated using a biomimetic approach, i.e. taking into account structural factors linked to the biology of plant fibres, such as the role of the cellular hollows and the presence of hierarchical levels, allowing the fibre to be built from the cell upwards. Fibers studied include celery and jute.

Key words: cellulosic fibres, biomimetics, plant fibres, celery, jute.

5.1 Biomimetic design of composite materials

Biomimetic materials design is a multidisciplinary approach, involving either the conception and realisation of new materials using biological tissues, with high environmental friendliness and a life cycle resulting in a lower resource depletion for the Earth, or tailoring materials for functions, as successfully addressed during evolution.[1]

However, mimicking a natural structure can mean different things in different situations. In other words, the extent of bio-inspiration in materials can vary, from copying the bare geometry of a natural object to reproducing its function or even providing a self-organisation paradigm for the material.[2, 3] The above approaches are in effect complementary, in the sense that the material is 'designed' in nature through self-organisation to perform the required function. Structurally, the inspiration from the natural structure can be obtained at super-molecular, molecular and sub-molecular level. In a more complex and vertical way, it is also possible to envisage that the material acquires higher levels of 'smartness' and subsequent functions, as far as it proceeds from the engineered material, perceived as a continuum, towards the biological material, containing a hierarchical structure, as represented in Fig. 5.1.

The biomimetic approach led in recent years to the conception of innovative fibre-reinforced composites, where at least the reinforcement phase, if not the

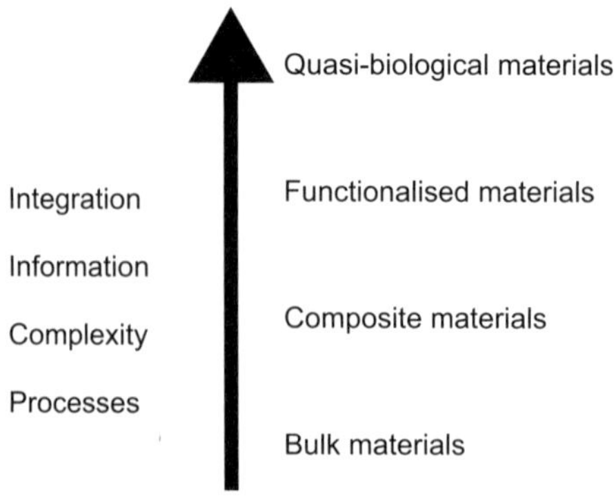

5.1 Vertical levels of complexity in materials (from reference 85).

whole composite, originated from nature. A fibrous morphology can be preferable for a number of reasons, including the possibility of embedding woven or knitted fibre tissues to form fracture-resistant composites: here, the inspiration can be from biological self-assembled tissues, not requiring energy-intensive or environmentally damaging processing steps.[4] Concentrating on the mechanical purpose of manufacturing composite, the matrix allows the re-distribution of concentrated loads and resistance to buckling.[5]

Potential materials for manufacturing innovative composites using a biomimetic approach are those related to cellulose fibres and those based on protein fibres. Five possibilities for reinforcement can be recognised: protein fibres from animals, genetically modified protein fibres, reconstituted animal cellulose (tunicin) fibres, cellulose fibres from agro-waste material, and cellulose or ligno-cellulose fibres extracted from plants. This chapter contains a review of these choices, but concentrates on the alternatives involving the use of cellulose fibres from plants.

General considerations on the possibilities of protein fibres as composites reinforcement are reported below. In particular, beta-keratin is an only partially hydrophilic material, which appears, at least in principle, ideal to be coupled with a hydrophobic matrix and/or with cellulose fibres as a hybrid composite.[6] A number of recent studies using keratin fibres from avian feathers as a reinforcement highlighted that the main issue to obtain a sufficient matrix–fibre compatibility would be having some control over the feather supply and possibly of their melanin content.[7–9]

Another possibility would be silk obtained from silkworm (*Bombyx mori*) and some spiders, such as *Nephila clavipes*. The former, in particular, has been used for centuries as a biomedical material for sutures, biodegradable and chemically suited to the function. Owing to some bio-compatibility problems to be ascribed to sericin contamination, its use as such has been discontinued. In recent years, bio-silk, obtained by genetically modifying silk proteins, has been introduced and this shows strong mechanical properties, making it suitable for use in composites.[10–11] The use of waste silk materials in composites is also reported in literature:[12] this is

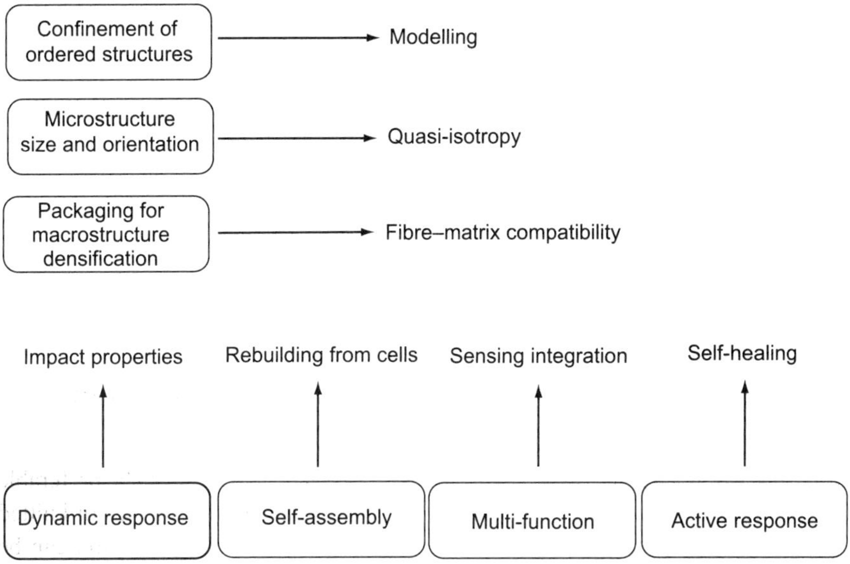

5.2 Mechanical design and functional design of a biomimetic composite.

of similar interest as obtaining fibres from agricultural waste, especially in that it may improve the life-cycle analysis (LCA) performance of the final material. Most frequently, agro-waste, such as rice husks, is used as a filler for thermoplastic polymers.[13] In some cases, however, it was also proposed as a material in (randomly oriented) fibrous form for materials reinforcement, although mainly applied to wooden board rather than fibreglass replacement.[14, 15]

5.2 Characteristics of biological materials in biocomposites

Biological materials offer a number of characteristics that are not always available in engineered materials. These characteristics should be available, at least in principle, when synthesising materials using biological sources, such as in composites including natural fibres as a reinforcement for a polymer matrix.

In addition to mechanical design properties, which suggest embedding fibres into a polymer matrix, composites, to be really biomimetic should be able to perform a number of functions, in what can be defined as functional design. These two levels of composite design are clarified, with reference to plant fibre composite, in Fig. 5.2. In more detail, the functions that biomimetic composites should have are:

- Building themselves in a hierarchical and optimised way, and, when necessary, coming back to the cell level to 'rebuild' the material (self-assembly). Routes for self-assembly, which would represent the final stage of an integrated process, including polymer synthesis and mineralisation, have only been proposed in silk replacement materials so far.[16]

- Performing different tasks when required (e.g. allow sensing integration). Here, chemical bio-sensing is, in principle, available in materials embedding cellulosic fibres by incorporation of glucose co-enzyme systems.[17] It appears to be more difficult to include mechanical sensing capabilities in the material. The main reason for this is owing to the reduction of mechanical properties induced by embedding fibre optics systems, usually in carbon fibre composites,[18] and this is unlikely to be compensated for by using cellulosic or lignocellulosic fibres.
- Responding in an active way to damage (e.g. self-healing). Self-healing of composite materials comprises two functions: enhanced damage visualisation and mechanical restoration through healing agent inserted in hollow fibres.[19] Usually borosilicate glass fibres have been used; however, the presence of hollow plant fibres would, at least in principle, allow them to be used for self-healing functions.

As discussed above, the potential of plant fibres in biomimetics depends on the role of the hollow in these fibres. A hollow fibre allows fluids and nutrients to diffuse in the plant. The presence of a hollow, or lumen, in a structure has been for quite some time regarded as solely detrimental for its mechanical resistance. However, it is worth pointing out that having a hollow in the centre of a supposedly cylindrical, loaded structure, means that the neutral axis in the structure is unloaded, a fact which may be beneficial from the point of view of shear tension in composites.[20]

The system of interest leading to the modelling/realisation/validation of plant fibre composites is reported in Fig. 5.3 relative to impact properties, which are important for possible high-tech use of these materials. The three aspects essential for the successful use of plant fibres in composites, i.e. fibre extraction, treatment and matrix coupling, can not be 'designed' without considering the presence of the fibre hollow. More generally, the difference between biological artefacts, such as plant fibres, and man-made ones, is particularly reflected in the presence of a hierarchical repeatability, which is different from the engineering macro-scale repeatability of components.[21] For fibre-reinforced composites, which are classically regarded as structures with a low order of hierarchical structure, fibres are embedded in a matrix to form anisotropic laminae that are, in turn, bonded together to form a laminate. Both fibres and matrix are considered to be continuous media in the analysis of the lamina, and the laminae are regarded as continuous in the analysis of the laminate, so that the degree of anisotropy would depend on the stacking sequence of laminae and the orientation of fibres within them.[22]

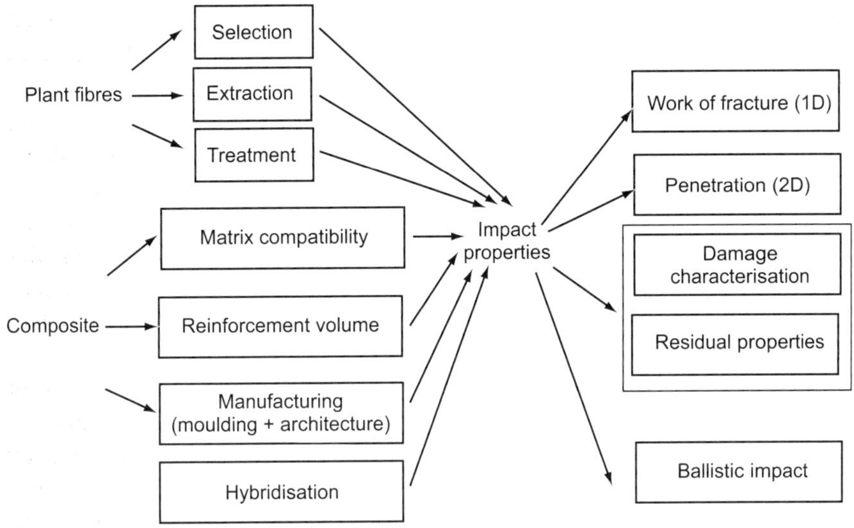

5.3 System of interest leading to impact-resistant structural plant-fibre composites.

5.3 Fibre extraction, fibre treatment and matrix compatibility in a biomimetic composite

Cellulose forms the structural component of plant fibres. Other components include hemicellulose, lignin and in some cases pectin and waxes, as well as some moisture content. Increasing as much as possible the cellulose percent in the final fibre would require fibre drying and elimination of pectins in the first place. In bast fibres, conventional field or dew retting, based on bacterial action, and followed by decortication, has a number of uncontrollable factors, which may affect the final quality of the fibre. This limitation is particularly significant when structural applications are envisaged for the fibre-reinforced laminates. In addition, it is not at all an environmentally friendly process, whose impact has been discussed in literature.[23, 24] There are also difficulties in applying binders to long fibre bundle lengths, since decortication technology is still not optimised in terms of mechanical reliability, separation efficiency and purity. To avoid extraction problems, an alternative is to leave the fibres unretted with the undesirable result of having more parenchyma, cuticle and epidermal tissue attached to the fibre tissue and therefore a weaker interface between the tissues.[25] For these reasons, on hemp fibres the possibility of green decortication and degumming technology has been explored,[26] whilst, on flax fibres, enzyme retting, involving the use of pectinase and galacturonase for pectins dissolution, is a quite well explored possibility.[27, 28]

Other fibres, which are obtained from fruit, such as coir from coconut, present a comparatively easier extraction, although retting by bacterial complexes is also commonly applied.[29] In this case, fibre length depends on the size of the de-husked coconut, which is highly variable in different plantations and even more in different regions.[30] A particular situation is that of bamboo, which, in spite of being a traditionally used material, especially in the building industry, was only rarely proposed in fibrous form as a reinforcement for polymeric matrices. In the few works existing on this subject, extraction was nevertheless carried out using a combination of mechanical (e.g. roll mill) and chemical methods.[31]

In general, it can be quite reasonably inferred that, in spite of extensive work on improving fibre extraction, the most common types of plant fibres used in composites have by no means sufficient properties for being used in structural components. As a consequence, it is not possible at this stage to rule out the use of fibre treatments to improve fibre characteristics. A large number of treatments have been proposed and applied, the significance of which has been discussed by Li *et al.*[32] In general, applied treatments tend in most cases to reproduce what has been realised on cotton fibres, in particular to obtain better dye tolerance for the tissue, a characteristic often termed dyeability. This is the case for sodium hydroxide, which proved effective in dissolving hemicellulose, forming microscopic sodium–cellulose local domains, depending on the diffusion of the alkali over the fibre surface.[33] The corresponding macroscopic effect comprises reducing fibre surface asperities and allowing quasi-cylindrical fibre sections to form.[34, 35] However, the improvement of fibre–matrix compatibility by geometrical effect appears to be compensated for by other negative chemical effects: for example mercerisation affects polymerisation, when cellulose fibres are cross-linked with polymer matrices, which can reduce the effectiveness of the treatment.[36]

In other instances, a better fibre–matrix compatibility has been achieved by adopting 'softer' treatments, such as silane, which was demonstrated to be effective, when working on glass fibres.[37–38] When concentrating on the need to have lower water absorption or anti-fungal protection of the fibres, the situation appears to be less defined, so that a number of treatments have been applied with some success. In practice, proposed treatments include acetylation,[39] a particularly flexible treatment, allowing e.g. the transformation of chitosan (i.e. polysaccharide) fibres into regenerated chitin (i.e. protein) fibres, an effective anti-fungal process;[40] benzoylation, particularly effective on cotton fibres in terms of elongation, moisture regain, friction, abrasion resistance and recovery;[41, 42] urea/microwaves;[43] isocyanate;[44] potassium permanganate/photocuring;[45] organic peroxide;[46] corona discharge and ultraviolet rays.[47]

In recent years, a matrix-specific treatment using maleic anhydride grafting on the polypropylene chain (MAPP) has often been applied, e.g. in references 48–50. This is a part of a present trend of employing polypropylene matrices, because they are thermoplastic and more easily recyclable, and as the base for obtaining co-mingled laminates, commonly used in the automotive industry and also proposed with plant fibres reinforcement, for better impact properties.[51]

Some of the above treatments, especially treatment with alkali, are simple and inexpensive, although their efficacy in improving fibre stiffness and fibre–matrix bond may not result in a similar increase of the static properties of the composite. In particular, it is necessary to optimise the treatment in terms of time, quantity or chemical and possibly temperature of application, verifying microscopically that the required surface modification has been obtained.[52] This optimisation process is by no means obvious and depends on the vegetable species involved and on a number of biological aspects, which will be also dealt with in Section 5.4.

An alternative method appears to be to increase the resistance of fibre/matrix interface by, e.g. allowing penetration of the polymeric resin into plant wall cells, a method demonstrated to be effective on flax fibre/epoxy resin composites.[53] In this regard, the real challenge appears to be improving fibre extraction to avoid the high cost and the not always sure efficacy of chemical treatment and to convert the extracted fibres into appropriate intermediate products as true 2D random mats, bi-directional and unidirectional.

In conclusion, it is still fair to say that no systematic or else optimised approach to fibre treatment is available so far, although a number of comparative studies exist in which different treatments are applied to the fibres, and materials properties obtained as a result are characterised.[54, 55] A recent review, although having a quite optimistic approach to fibre treatment as necessary and potentially able to substantially increase both plant fibres strength and fibre–matrix compatibility, could not neglect the fact that in real terms these issues are far from resolved.[56]

5.4 Approaches to the realisation of plant fibre composites

5.4.1 Possibilities for plant fibre selection in composites

Plant fibres are currently selected in composites, mainly according to geo-political and economical aspects. As a consequence, a number of plant fibres have been proposed in the last decades for reinforcement of polymeric matrices. A list, which is by no means guaranteed exhaustive, is presented in Table 5.1. From an economic point of view, it is significant to notice whether the fibre production represents the main reason for that cultivation, or else it is only a marginal product among a great number of other products. For example, switchgrass can be used also for fibre production, although it is almost exclusively cropped for bio-fuel production, while on the other side there are fibres whose production belongs to a very large and complex industrial sector, such is the case with coir and the coconut industry. Most cases are intermediate between these two indeed.

Of course, the transition to a selection more based on materials *vs.* properties considerations, such as in an Ashby diagram, which exist already in the more general field of natural materials,[57] would require a large number of comparative studies between composites, obtained by the same manufacturing procedure using different plant fibres as a reinforcement. Some of these studies do exist, as reported

Table 5.1 Plant fibres used or proposed for use for materials reinforcement

Plant	Botanic name	Fibres extracted from
Abaca	*Musa* textiles	Leaf
Banana	*Musa sapientum*	Leaf
Bamboo	Various species	Stem
Betelnut	*Araca catechu*[86]	Seed hair
Celery	*Apium graveolens*[87]	Stem
Coir	*Cocos nucifera*	Fruit hair
Date palm	*Phoenix dactylifora, Phoenix sylvestris*	Leaf base (netted structure)
Esparto	*Lygeum spartum, Stipa tenacissima*	Stem
Flax	*Linum usitatissimum*	Stem
Hemp	*Cannabis sativa*	Stem
Henequen	*Agave fourcroydes*	Leaf
Indian grass	*Sorghastrum nutans*[88]	Stem
Jute	*Corchorus* sp.	Stem
Kapok	*Ceiba pentandra, Ceiba occidentalis*	Fruit hair
Kenaf	*Hibiscus cannabinus*	Stem
Lady's fingers	*Abelmoschus esculentus*[88]	Bark
New Zealand flax	*Phormium tenax*[89]	Stem
Oil palm	*Elaeis guineensis*	Fruit hair
Palmyra	*Borassus* sp.[90]	Leaf
Piassava	*Attalea funifera*	Leaf
Pineapple	*Ananas comosus*	Leaf
Ramie	*Boehmeria nivea*	Stem
Roselle	*Hibiscus sabdariffa*	Stem
Royal palm	*Roystonea regia*	Leaf
Sisal	*Agave sisalana*	Leaf
Spanish Broom	*Spartium junceum*[91]	Stem
Sunn hemp	*Crorolaria juncea*	Stem
Switchgrass	*Panicum virgatum* L.[92]	Stem
Talipot	*Corypha umbraculifera*[90]	Leaf
Vetiver	*Vetiveria zizanoides*[93]	Stem

Note: references are given for fibres not typically used in composites

in Section 5.3 for fibre treatment, also on particular aspects of fibre performance, such as impact properties,[58] but a general database is still missing and, in a sense, well overdue.

One reason for this lack of large comparative studies is the complexity of factors involved, which is especially critical when biological factors are included, such as fibres maturity, taxonomic differences, etc., which can largely affect the final properties of the composite produced. In recent years, these aspects tend to be more recognised, so that composite studies tend to be more accurate in declaring the origin of crops from which fibres are extracted: this used to be generally overlooked, leading to the difficulty of tracing back exactly to the non-local species. A good example of this strategy is given by Faria *et al.*,[59] where the study leading to banana fibre/polypropylene composites is more dedicated to the microscopic characterisation of fibres, reflected macroscopically, e.g. in the degree of crystallinity and the lignin content, than to the crude improvement of fibre strength with an appropriate treatment. Chemical treatment (MAPP in this case) is applied only once a sound knowledge of fibre properties is acquired: this might be a first step towards fibre selection.

A second reason, which appears to be more critical for the possible success of plant fibres as a reinforcement in composites may be deduced. Including biological structures, such as plant fibres, in man-made materials would require changing the engineering perspective to them. In practice, this would mean leaving aside the traditional approach to composites, as formed by fibres, embedded in a matrix with some kind of, possibly strong, interface. This approach requires that the laminate has a recognisable stacking sequence, to which the composite owes its quasi-isotropic behaviour. It has also been proposed as the basis for the production of 'biomimetic composite laminates' reproducing in their stacking structure the one used in biological composites, e.g. insect cuticles and fish scales. The bare reproduction of biological stacking sequences led to interesting results in terms of improvement of mechanical and impact absorption properties in carbon fibre reinforced composites.[60] The application of this philosophy to plant fibre composites would require nonetheless a greater level of complexity to be applied, introducing other factors: in particular, bamboo fibre has been described on a micro scale as a helical, multi-layered hollow cylinder.[61] Accepting these three characteristics of the reinforcement can be sufficient for some general modelling. However, as a consequence of the large dimensional variability of the fibres, starting at the lowest level, from the cell, as shown for example in Fig. 5.4a, representing a celery fibre structure, and in Fig. 5.4b, representing a jute fibre structure, every single laminate would have its personal 'stacking sequence' and hardly predictable properties. In other words, it can be deemed necessary to accept that some form of hierarchical structure is present in the composite, which is incidentally a fundamental observation of studies on natural composite laminates, such as e.g. arthropod cuticles.[62]

Coming back to fibre selection, the above reasons make the traditional approach insufficient for allowing the comparison between different plant fibres as a reinforcement; the different origin of the plant structure (bast, leaves, seed, fruits) would require also the interface and the stacking sequence to be modified to account for these differences. In contrast, the biomimetic approach may allow fibres selection on an objective basis, considering both the presence of hollow and their helical structure, formed by stronger (or crystalline) and weaker (or amorphous) parts. It is worthy to note nonetheless that this modelling can work if reasonably accurate predictions over the influence of defects are obtained. In real plant fibre composites, defects and dimensional variations are easily observable at any level of the hierarchical structure, as shown in Fig. 5.4a (celery fibre embedded in epoxy matrix) and Fig. 5.4b (bundles of jute fibres embedded in polyester matrix). The influence of defects in these materials has also obtained some

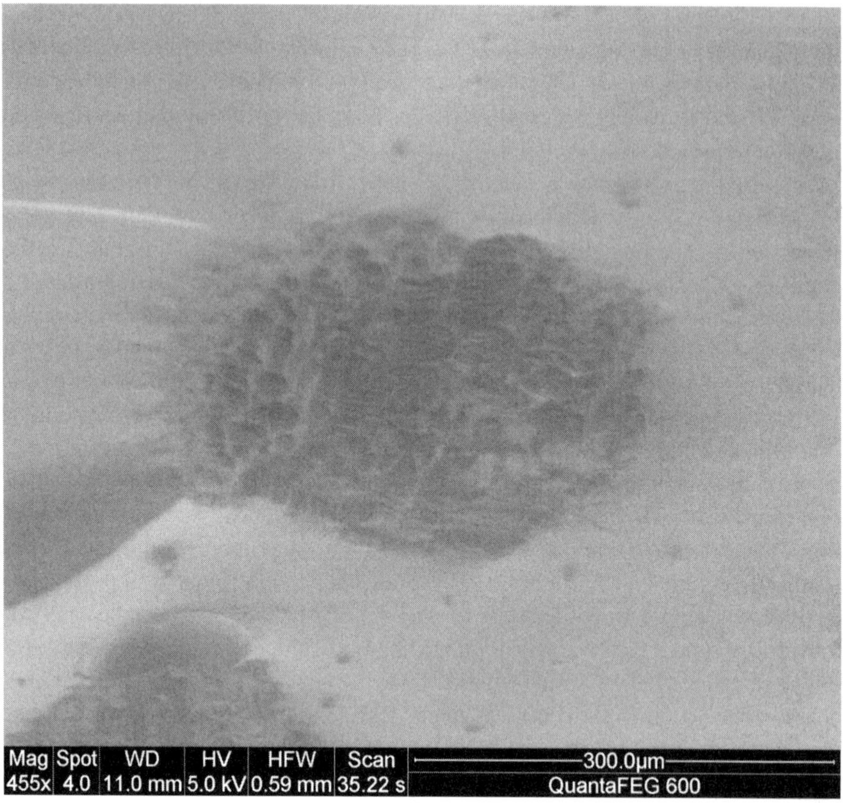

5.4a Structure of a celery fibre.

attention: it is well known e.g. that, in plant fibre composites, the longer the clamping length the lower the tensile strength; this is because of the presence of a larger number of fibre defects, i.e. failure initiators.[63] A similar effect was observed in impact-damaged plant fibre composites, when areas showing poor adhesion, which occur because of geometrical and structural variations of the fibre bundles, act as triggers for impact failure.[64]

5.4.2 Life-cycle analysis of plant fibre composites and hybridisation as an environmental trade-off

The position of LCA in a typical economic system comprising the use of plant fibres is given in Fig. 5.5. More details on the general environmental significance (and true possibility) of replacing, e.g. glass fibre composites with plant fibre composites, as results from LCA, are given in Joshi *et al.*[65] The main environmental advantages of plant fibre over glass fibre composite appeared to correlate to the lower use of fossil fuels, even considering the use of fertilisers in agriculture, and

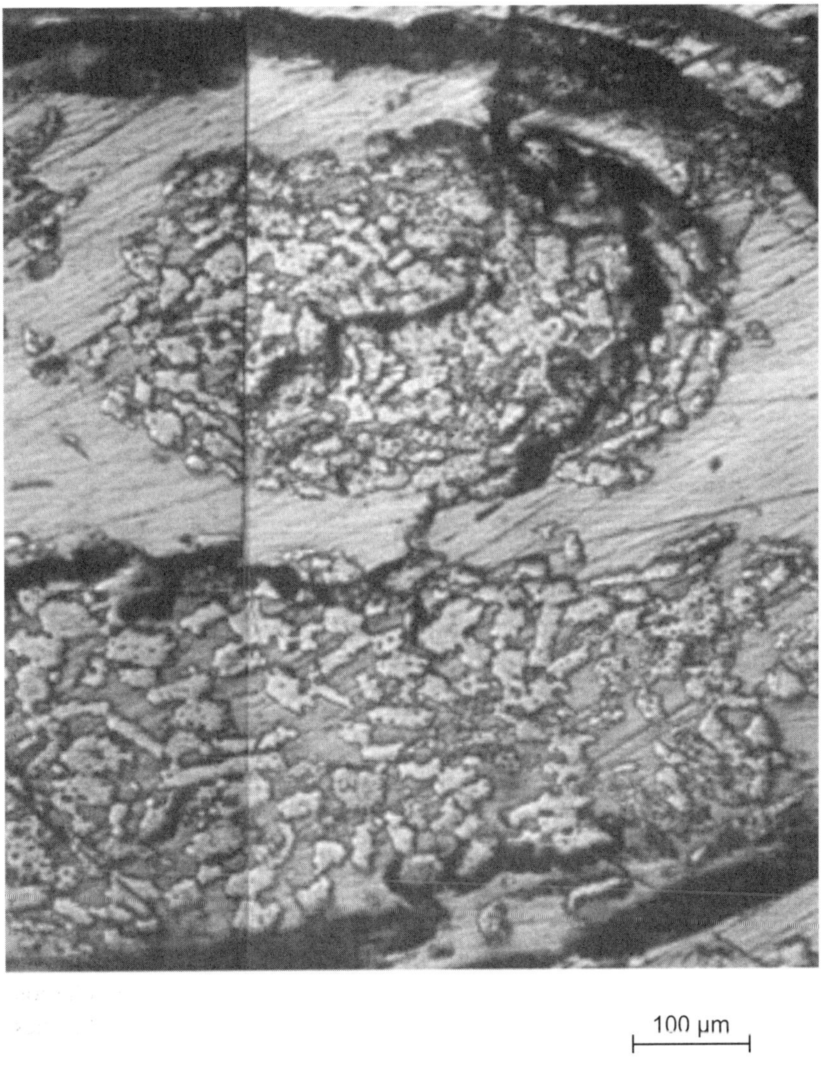

5.4b Structure of jute fibre bundles.

the lower level of eutrophication, as recognised also elsewhere.[66] A number of caveats can nonetheless be raised: first, the effect on eutrophication would need an absolute control over all the steps of agricultural operations, such as retting, which does not seem to be the case so far;[67] second, LCA results appear to be strongly influenced by the envisaged application, especially in the sense that the quantity of chemicals applied for treatment is depending on the level of improvement of fibre properties required.[68] Of course, selection of the fibres more suited to the application, albeit not easily obtainable, could assist in reducing the need for chemical treatment.

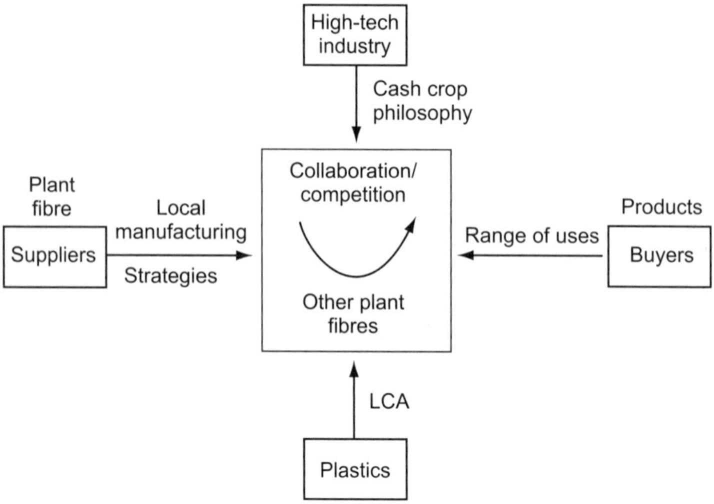

5.5 General economic system of the production of plant fibre composites.

In fact, in more recent studies LCA comprises three parts: inventory, dynamics, and validation, although traditionally the latter is omitted or disregarded.[69] For LCA studies involving natural fibres, validation cannot be neglected, but it has been hardly investigated so far. In particular, it is essential that the sustainability of the whole manufacturing procedure is assessed, which might involve a number of processes, such as fibre treatment and extraction, far from optimised, even if reasonably environmental friendly when compared with glass fibre sizing. In addition, the selection of the best fibre for a given application may represent a necessary development, which will possibly include modelling the material with analogies appropriate with the use of natural fibres. This 'natural modelling' would particularly require, as mentioned before, accounting for the helical symmetry of stacking sequences, the presence of hollows and the role of water and/or relative humidity in materials processing. Some examples of natural modelling are starting to be considered in composites, as is the case with the development of a particular synthetic fibre, inspired to a biological hollow fibre, referred to as 'technical plant stem'.[70] This can potentially lead to the active incorporattion of defects, for example with probabilistic methods, into modelling of natural fibre composites, in a way not dissimilar to what is done with dislocations in metals.

Meanwhile, the inherent difficulty of validating data obtained from LCA, considering materials with large variability, such as biological fibres, would in some cases require an intermediate approach: hybridisation. It has been recognised that hybrids have often been used in materials science and technology to expand design space, allowing a material to be designed with specific properties.[71] In a specific case, however, a hybrid laminate including two species of fibres, of which at least one is extracted from a plant, albeit slightly complicated in manufacturing terms, may act environmentally in a kind of trade-off, offering some environmental gain with respect to man-made composites. It could also be 'designed for function', provided of course the assumptions made on it are as close to reality as possible.

In practice, hybridisation, although largely used over the last two decades on E-glass/carbon fibres,[72] kevlar/S-glass[73] and also E-glass/plant fibres[74, 75] with some limited success, has recently been re-introduced in a more global sense in composites reinforced with some plant fibres, in particular the more lignified ones, such as jute and sisal.[76, 77] However, most studies here again are concerned with finding the most suitable stacking sequence (usually plant fibres are used for inner layers and glass fibres for outer layers). This approach can be useful especially for the sake of simplicity, in order to evaluate if a positive hybridisation effect is reached, with reference to the possible validity of the 'rule of mixtures' for static mechanical properties.[78] There is no reason, nonetheless, that hybrids could not be obtained with a more complex structure, with a higher hierarchical order. In practice, as mentioned above, the reduction of mechanical and structural properties obtained with the introduction of some amount of plant fibres can be compensated by some environmental benefits, due to the higher grade of biodegradability of the material. Both the optimal amount of fibres to be introduced and the predicted environmental gain are strongly dependent on the application envisaged for the material. In Table 5.2, a small number of recent studies on hybrid configurations including natural fibres are presented and briefly discussed to give an idea of the issues involved in hybridisation.

Hybrids present advantages that are perfectly suitable in a biomimetic approach: the need, e.g. to account for the non-avoidable presence of water, allow hybrids to be employed in applications for which biological materials are in principle excluded, such as for dielectrics. Hybrids including plant fibres with higher lignin content, such as palm oil fibres embedded in natural rubber matrices as bio-dielectrics, have been proposed to reduce the environmental impact of these materials.[79] It is interesting to note that a trend towards extending the application of hybridisation between fibres exists, for example with hybrids including partially hydrophilic fibres, such as those obtained from avian feathers. Where, however, the need for extensive chemical treatment exists (sanitisation), this represents another source of environmental impact thus modifying the LCA data of the final material.[80, 81]

5.5 Conclusions: plant fibre selection for composites reinforcement

A biomimetic approach to the production of composites including plant fibres represents a possibility for addressing the problem of obtaining a more sustainable material, with a more acceptable LCA profile. In practice, the two combined aspects of accepting the unpredictable presence of defects and seeing the

Table 5.2 Some examples of hybrids including natural fibres

Hybrid fibres	Advantages	Limits
Short glass/empty fruit bunch[94]	Introduced in a composite waste material, reasonably clean, generally improving LCA	Considerable dimensional variations. Difficult compatibility
Jute/glass[76]	Good for industrial applications. Well-known hybrid for decades. Could revive jute market	Quite far from a biomimetic material. Traditional composite modelling (classical laminate). Could require aggressive treatments
Sisal/oil palm[79]	Plant/plant hybrids allow optimising properties of natural fibre composites. Closer to a biomimetic material	Two plant fibres mean double degree of variability in properties
Chicken feathers/ aspen wood[80]	Introduced in a composite waste material, to be treated, ideally improving LCA	Chemical treatment required. Difficult compatibility. Hard quality control of feather supply
Sisal/glass[77]	Ideal for industrial applications. Applications could match the traditional ones for composites (e.g. automotive panels)	Quite far from a biomimetic material. Traditional composite modelling (classical laminate)

opportunities offered by hollow fibres can lead to a new material modelling that includes the development of composite materials different from traditional ones, with some form of hierarchical structure. It is worthy of note that fibre extraction through biological retting can be designed to recognise the degree of hierarchical order present in the plant structure from the cell up to the fibre bundle and therefore contribute to materials design.[82] An essential aspect of this evolution for composite materials is the possibility of constituting woven tissues, which will be more similar to spontaneous tissues formed by plant fibres, according to their respective structure and geometry. In terms of materials selection, this would broaden the possibility of choosing plant fibres for composites to unusual or even infesting crops, such as e.g. nettle.[83]

In environmental terms, it is possibly fair to assume that a biomimetic approach to composites would allow using plant fibres in more sustainable materials, provided that the repeatability of the microstructure (hierarchical nature of the material) would lead to a repeatable result, even if it is not easy to identify it for materials modelling.

However, there are other aspects which need to be clarified, before hierarchically ordered and environmentally sustainable composites are available; in particular, environmental advantages need not be nullified by the presence of other sensitive issues. This applies particularly to two types of reinforcement recently proposed for composites: bio-silk fibres, where bio-compatibility for envisaged medical use requires sericin removal, obtained by genetic modification,[11] and keratin fibres from avian feathers, where sanitisation, usually by sodium hypochlorite, is required to achieve fibres suitable for use.[84]

An important conclusive consideration is in regard to the use of hybridisation procedures for composite manufacturing: these not only can suggest new applications for composites and cover gaps in existing properties, but also improve the environmental sustainability of existing materials, such as fibreglass with the introduction of plant fibres. Hybridisation, therefore, needs to be considered in the most general possible sense, looking not only at the compatibility between the originating materials, but also at their respective properties, such as in some of the examples discussed in Table 5.2.

5.6 References

1 Santulli C, Biomimetic interest and possibilities of plant fibres for the production of biodegradable impact-resistant composites, Proceedings 5th Summer School on Materials Science, Jesi, August 2003, pp. 119–136.

2 Vincent JFV, Borrowing the best from Nature. In *Encyclopaedia Britannica yearbook*, p. 169, 1995 (Bettman Archive).

3 Imai H, Oaki Y and Kotachi A, A biomimetic approach for hierarchically structured inorganic crystals through self-organization, *Bulletin of the Chemical Society of Japan* **79**, 2006, 1834–1851.

4 Viney C, Self-assembly to fibrous materials: concepts, opportunities and challenges, *Current Opinion in Solid State and Materials Science* **8**, 2004, 95–101.

5 Hull D, Matrix dominated properties of polymer matrix composite-materials, *Materials Science and Engineering A* **184**, 1994, 173–183

6 Barone JR and Schmidt WF, Polyethylene reinforced with keratin fibers obtained from chicken feathers, *Composites Science and Technology* **65**, 2005, 173–181.

7 Barone JR, Schmidt WF and Liebner CFE, Compounding and molding of polyethylene composites reinforced with keratin feather fiber, *Composites Science and Technology* **65**, 2005, 683–692.

8 Cameron GJ, Wess TJ and Bonser RHC, Young's modulus varies with differential orientation of keratin in feathers, *Journal of Structural Biology* **143**, 2003, 118–123.

9 Kavitha A, Boopalan K, Radhakrishnan G, Sankaran S, Das BN and Sastry TP, Preparation of feather keratin hydrolyzate–gelatin composites and their graft copolymers, *Journal of Macromolecular Science – Pure and Applied Chemistry* **A42**, 2005, 1703–1713.

10 Sang ML, Cho D, Won HP, Seung GL, Seong OH and Drzal LT, Novel silk/poly(butylene succinate) biocomposites: the effect of short fibre content on their mechanical and thermal properties, *Composites Science and Technology* **65**, 2005, 647–657.

11 Altman GH, Diaz F, Jakuba C, Calabro T, Horan RL, Chen J, Lu H, Richmond J and Kaplan DL, 2003, Silk-based biomaterials, *Biomaterials* **24**, 401–416.

12 Priya SP, Ramakrishna HV and Rai KS, Utilization of waste silk fabric as reinforcement in epoxy phenol cashew nut shell liquid toughened epoxy resin: studies on mechanical properties, *Journal of Composite Materials* **40**, 2006, 1301–1311.

13 Bera J and Kale DD, Effect of particle size of chemically treated rice husk powder on the properties of filled polypropylene, *Journal of Polymer Materials* **23**, 2006, 203–209.

14 Mishra S and Naik JB, Absorption of water at ambient temperature and steam in wood–polymer composites prepared from agrowaste and polystyrene, *Journal of Applied Polymer Science* **68**, 1998, 681–686.

15 Samal RK, Giri G and Bhuyan BL, Chemical modification of lignocellulosic fibers II – functionality changes and graft-copolymerization of methyl methacrylate onto pineapple leaf fibers, *Journal of Polymer Materials* **11**, 1994, 113–119.

16 O'Brien JP, Fahnestock SR, Termonia Y and Gardner KCH, Nylons from nature: synthetic analogs to spider silk, *Advanced Materials* **10**, 1998, 1185–1195.

17 Morales A, Céspedes F and Alegret S, Graphite–methacrylate biocomposite material with renewable sensing surface for reagentless amperometric biosensors based on glucose dehydrogenase, *Materials Science and Engineering C* **7**, 2000, 99–104.

18 Shin CS and Chiang CC, Fatigue damage monitoring in polymeric composites using multiple fiber Bragg gratings, *International Journal of Fatigue* **28**, 2006, 1315–1321.

19 Pang JWC and Bond IP, Bleeding composites: damage detection and self-repair using a biomimetic approach, *Composites Part A* **36**, 2005, 183–188.

20 Cannon BJ and Brei D, Feasibility study of microfabrication by coextrusion (MFCX) hollow fibers for active composites, *Journal of Intelligent Material Systems and Structures* **11**, 2000, 659–670.

21 Lakes R, Materials with structural hierarchy, *Nature* **361**, 1993, 511–515.

22 Jones RM, *Mechanics of Composite Materials*, Taylor and Francis, 1998, ISBN-13 978-1560327127.

23 Anthony WS, Separation of fiber from seed flax straw, *Applied Engineering in Agriculture* **18**, 2002, 227–233.

24 Gratton JL and Chen Y, Development of a field-going unit to separate fiber from hemp (*Cannabis sativa*) stalk, *Applied Engineering in Agriculture* **20**, 2004, 139–145.

25 Mustata A, Avram D and Slabu V, Using unretted flax to obtain yarns and woven fabrics, *Fibres and Textiles in Eastern Europe* **14**, 2006, 29–32.

26 Riddlestone S, Stott E, Blackburn K and Brighton J, A technical and economic feasibility study of green decortication of hemp fibre for textile uses, *Journal of Industrial Hemp* **11**, 2006, 25–55.

27 Akin DE, Foulk JA, Dodd RB and McAlister DD, Enzyme-retting of flax and characterization of processed fibers, *Journal of Biotechnology* **89**, 2001, 193–203.

28 Zhang J, Henriksson H, Szabo IJ, Henriksson G, Johannson G, The active component in the flax-retting system of the zygomycete *Rhizopus otyzae* sb is a family 28 polygalacturonase, *Journal of Industrial Microbiology and Biotechnology* **32**, 2005, 431–438.

29 Ravindranath AD and Bhosle S, Development of a bacterial consortium for coir retting, *Journal of Scientific and Industrial Research* **59**, 2000, 140–143.

30 Konduru S, Evans MR and Stamps RH, Coconut husk and processing effects on chemical and physical properties of coconut coir dust, *Hortscience* **34**, 1999, 88–90.

31 Deshpande AP, Rao MB and Rao CL, Extraction of bamboo fibers and their use as reinforcement in polymeric composites, *Journal of Applied Polymer Science* **76**, 2000, 83–92.

32 Li X, Tabil LG and Panigrahi S, Chemical treatments of natural fiber for use in natural

fiber-reinforced composites: a review, *Journal of Polymers and the Environment* **15**, 2007, 25–33.
33 Schoeck J, Davies RJ, Martel A and Riekel C, Na–cellulose formation in a single cotton fiber studied by synchrotron radiation microdiffraction, *Biomacromolecules* **8**, 2007, 602–610.
34 Bledzki AK, Fink HP and Specht K, Unidirectional hemp and flax EP- and PP-composites: influence of defined fiber treatments, *Journal of Applied Polymer Science* **93**, 2004, 2150–2156.
35 Fink HP, Walenta E and Kunze J, The structure of natural cellulosic fibres – Part 2. The supermolecular structure of bast fibres and their changes by mercerization as revealed by x-ray diffraction and C-13 NMR spectroscopy, *Papier* **53**, 1999, 534–542.
36 Zhou LM, Yeung KW, Yuen CWM and Zhou X, Characterization of ramie yarn treated with sodium hydroxide and crosslinked by 1,2,3,4-butanetetracarboxylic acid, *Journal of Applied Polymer Science* **91**, 2004, 1857–1864.
37 Girones J, Mendez JA, Boufi S, Vilaseca F and Mutje P, Effect of silane coupling agents on the properties of pine fibers/polypropylene composites, *Journal of Applied Polymer Science* **103**, 2007, 3706–3717.
38 Yun SH, Cho DH, Kim J, Lim S, Lee GW, Park M and Lee S, Effect of silane coupling agents with different organo-functional groups on the interfacial shear strength of glass fiber/Nylon 6 composites, *Journal of Materials Science Letters* **22**, 2003, 1591–1594.
39 Rana AK, Basak RK, Mitra BC, Lawther M and Banerjee AN, Studies of acetylation of jute using simplified procedure and its characterization, *Journal of Applied Polymer Science* **64**, 1997, 1517–1523.
40 East GC and Yiming Q, Wet spinning of chitosan and the acelylation of chitosan fibers, *Journal of Applied Polymer Science* **50**, 1993, 1773–1779.
41 Ferreira FC, Curvelo AAS and Mattoso LHC, Preparation and characterisation of benzylated sisal fibers, *Journal of Applied Polymer Science* **89**, 2003, 2957–2965.
42 Lorand EJ and Georgi EA, The mechanism of cellulose benzylation, *Journal of the American Chemical Society* **59**, 1937, 1166–1170.
43 Sun H, Lin L, Jiang X and Bai X, The improvement of dyeability of flax fibre by microwave treatment, *Pigment and Resin Technology* **34**, 2005, 190–196.
44 Ganan P and Mondragon I, Effect of fiber treatments on mechanical behavior of short fique fiber-reinforced polyacetal composites, *Journal of Composite Materials* **39**, 2005, 633–646.
45 Khan MA, Hassan MM, Taslima R and Mustafa AI, Role of pretreatment with potassium permanganate and urea on mechanical and degradable properties of photocured coir (*Cocos nucifera*) fiber with 1,6-hexanediol diacrylate, *Journal of Applied Polymer Science* **100**, 2006, 4361–4368.
46 Arzondo LM, Perez CJ and Carella JM, Injection molding of long sisal fiber-reinforced polypropylene: effects of compatibilizer concentration and viscosity on fiber adhesion and thermal degradation, *Polymer Engineering and Science* **45**, 2005, 613–621.
47 Gassan J and Gutowski VS, Effects of corona discharge and UV treatment on the properties of jute-fibre epoxy composites, *Composites Science and Technology* **60**, 2000, 2857–2863.
48 Vallejos ME, Girones J, Vilaseca F, Mendez JA, Lopez A and Mutje P, Polypropylene reinforced with hemp strands: study of the influence of propylene modified with maleic anhydride as coupling agent, *Afinidad* **63**, 2006, 94–105.
49 Park JM, Quang ST, Hwang BS and DeVries KL, Interfacial evaluation of modified jute and hemp fibers/polypropylene (PP)–maleic anhydride polypropylene copolymers

(PP–MAPP) composites using micromechanical technique and nondestructive acoustic emission, *Composites Science and Technology* **66**, 2006, 2686–2699.

50 Mohanty S, Verma SK, Nayak SK and Tripathy SS, Influence of fiber treatment on the performance of sisal–polypropylene composites, *Journal of Applied Polymer Science* **94**, 2004, 1336–1345.

51 Goutianos S and Peijs T, The optimisation of flax fibre yarns for the development of high-performance natural fibre composites, *Advanced Composites Letters* **12**, 2003, 237–241.

52 Cao Y, Shibata S and Fukumoto I, Mechanical properties of biodegradable composites reinforced with bagasse fibre before and after alkali treatments, *Composites Part A* **37**, 2006, 423–429.

53 Hepworth DG, Bruce DM, Vincent JFV and Jeronimidis G, The manufacture and mechanical testing of thermosetting natural fibre composites, *Journal of Materials Science* **35**, 2000, 293–298.

54 Joseph K, Thomas S and Pavithran C, Effect of chemical treatment on the tensile properties of short sisal fibre-reinforced polyethylene composites, *Polymer* **37**, 1996, 5139–5149.

55 Pothan LA, George J and Thomas S, Effect of fiber surface treatments on the fiber–matrix interaction in banana fiber reinforced polyester composites, *Composite Interfaces* **9**, 2002, 335–353.

56 Li X, Tabil LG and Panigrahi S, Chemical treatments of natural fiber for use in natural fiber-reinforced composites: a review, *Journal of Polymers and the Environment* **15**, 2007, 25–33.

57 Wegst UGK and Ashby MF, The mechanical efficiency of natural materials, *Philosophical Magazine* **84**, 2004, 2167–2181.

58 Wambua P, Ivens J and Verpoest I, Natural fibres: can they replace glass in fibre reinforced plastics? *Composite Science and Technology* **63**, 2003, 1259–1264.

59 Faria H, Cordeiro N, Belgacem MN and Dufresne A, Dwarf cavendish as a source of natural fibers in poly(propylene)-based composites, *Macromolecular Materials and Engineering* **291**, 2006, 16–26.

60 Tanimoto T, Damage tolerant behaviors of a biomimetic CFRP laminate, *Advanced Composite Materials* **11**, 2002, 11–19.

61 Li SH, Zhang RH, Fu SY, Chen X, Zhou BL and Zeng QYA, Biomimetic model for fiber-reinforced composite materials, *Journal of Materials Science and Technology (China) (USA)* **10**, 1994, 34–38.

62 Vincent JFV, Arthropod cuticle: a natural composite shell system, *Composites Part A* **33**, 2002, 1311–1315.

63 Nechwatal A, Mieck KP and Reussmann T, Developments in the characterization of natural fibre properties and in the use of natural fibres for composites, *Composites Science and Technology* **63**, 2003, 1273–1279.

64 Santulli C, Janssen M and Jeronimidis G, Partial replacement of E-glass fibres with flax fibres in composites and effect on falling weight impact performance, *Journal of Materials Science* **40**, 2005, 3581–3585.

65 Joshi SV, Drzal LT, Mohanty AK and Arora S, 2004, Are natural fiber composites environmentally superior to glass fiber reinforced composites?, *Composites Part A* **35**, 371–376.

66 Corbiere-Nicollier T, Gfeller-Laban B, Lundquist L, Leterrier Y, Manson JAE and Jolliet O, Life cycle assessment of biofibres replacing glass fibres as reinforcement in plastics, *Resources Conservation and Recycling* **33**, 2001, 267–287.

67 Ciroth A and Becker H, Validation – the missing link in life cycle assessment – towards pragmatic LCAs, *International Journal of Life Cycle Assessment* **11**, 2006, 295–297.

68 Dahllöf L and Steen B, A statistical approach for estimation of process flow data from production of chemicals of fossil origin, *International Journal of Life Cycle Assessment* **12**, 2007, 103–108.

69 Rosentrater KA and Otieno AW, Considerations for manufacturing bio-based plastic products, *Journal of Polymers and the Environment* **14**, 2006, 335–346.

70 Milwich M, Speck T, Stegmaier T and Planck H, Biomimetics and technical textiles: solving engineering problems with the help of nature's wisdom, *American Journal of Botany* **93**, 2006 1455–1465.

71 Ashby MF, Hybrids to fill holes in material property space, *Philosophical Magazine* **85**, 2005, 3235–3257.

72 Sonparote PW and Lakkad SC, Mechanical properties of carbon glass-fiber reinforced hybrids, *Fibre Science and Technology* **16**, 1982, 309–312.

73 Qiu YP and Schwartz P, Micromechanical behavior of Kevlar-149/S-glass hybrid 7-fiber microcomposites. 2. Stochastic modeling of stress-rupture of hybrid composites, *Composites Science and Technology* **47**, 1993, 303–315.

74 Clark RA and Ansell MP, Jute and glass-fiber hybrid laminates, *Journal of Materials Science* **21**, 1986, 269–276.

75 Pavithran C, Mukherjee PS, Brahmakumar M and Damodaran AD, Impact properties of sisal-glass hybrid laminates, *Journal of Materials Science* **26**, 1991, 455–459.

76 Ahmed KS and Vijayarangan S, Elastic property evaluation of jute-glass fibre hybrid composite using experimental and CLT approach, *Indian Journal of Engineering and Materials Science* **13**, 2006, 435–442.

77 John K and Naidu SV, Sisal fibre/glass fibre hybrid composites: the impact and compressive properties, *Journal of Reinforced Plastics and Composites* **23**, 2004, 1253–1258.

78 Marom G, Fisher S, Tuler FR and Wagner HD, Hybrid effects in composites: conditions for positive or negative effects versus rule of mixtures behaviour, *Journal of Materials Science* **13**, 1978, 1419–1426.

79 Jacob M, Varughese KT and Thomas S, Dielectric characteristics of sisal-oil palm hybrid biofibre reinforced natural rubber biocomposites, *Journal of Materials Science* **41**, 2006, 5538–5547.

80 Winandy JE, Muehl JH and Micales JA, Potential of chicken feather fibre in wood MDF composites, Proceedings EcoComp 2003, Queen Mary University of London, 1–2 September 2003. Paper P20.

81 Martinez-Hernandez AL, Velasco-Santos C, De-Icaza M and Castano VM, Dynamical–mechanical and thermal analysis of polymeric composites reinforced with keratin biofibers from chicken feathers, *Composites Part B* **38**, 2007, 405–410.

82 Ganan P, Zuluaga R, Velez JM and Mondragon I, Biological natural retting for determining the hierarchical structuration of banana fibers, *Macromolecular Bioscience* **4**, 2004, 978–983.

83 Vogl CR and Hartl A, Production and processing of organically grown fiber nettle (*Urtica dioica* L.) and its potential use in the natural textile industry: a review, *American Journal of Alternative Agriculture* **18,** 2003, 119–128.

84 Rijavec T and Bukosek V, Achievements in the development of fibres for technical applications, *Tekstil* **53**, 2004, 562–575.

85 Langella C, Nuovi paesaggi materici. Design e tecnologia dei materiali, Alinea Editrice, Firenze, 2003, ISBN 8881257130, p. 39.

86 Bora MN, Talukdar C and Talukdar M, Tensile properties of plant fibres readily available in north-eastern region of India, *Indian Journal of Fibres and Textile Research* **24**, 1999, 172–176.

87 Caneva C, De Rosa IM, Santulli C and Sarasini F, Reinforcement effect of untreated celery fibres on epoxy matrix composites, *International Journal of Materials and Product Technology*, 2007, in press.

88 Liu WJ, Mohanty AK, Askeland P, Drzal LT and Misra M, Influence of fiber surface treatment on properties of Indian grass fiber reinforced soy protein based biocomposites, *Polymer* **45**, 2004, 7589–7596.

89 Carr DJ, Cruthers NM, Laing RM and Niven BE, Fibers from three cultivars of New Zealand flax (*Phormium tenax*), *Textile Research Journal* **75**, 2005, 93–98.

90 Satyanarayana KG, Ravikumar MK, Sukumaran K, Mukherjee PS, Pillai SGK and Kulkarni AG, Structure and properties of some vegetable fibers. 3. Talipot and palmyrah fibres, *Journal of Materials Science* **21**, 1986, 57–63.

91 Angelini LG, Lazzeri A, Levita G, Fontanelli D and Bozzi C, Ramie (*Boehmeria nivea* L. Gaud.) and Spanish Broom (*Spartium junceum* L.) fibres for composite materials: agronomical aspects, morphology and mechanical properties, *Industrial Crops and Products* **11**, 2000, 145–161.

92 Radiotis T, Li J, Goel K and Eisner R, Fiber characteristics, pulpability, and bleachability of switchgrass, *Tappi Journal* **82**, 1999, 100–105.

93 Ruksakulpiwat Y, Suppakarn N, Sutapun W and Thomthong W, Vetiver–polypropylene composites: physical and mechanical properties, *Composites Part A* **38**, 2007, 590–601.

94 Anuar H, Ahmad SH, Rasid R and Daud NSN, Tensile and impact properties of thermoplastic natural rubber reinforced short glass fiber and empty fruit bunch hybrid composites, *Polymer–Plastics Technology and Engineering* **45**, 2006, 1059–1063.

Part II

Biomimetic applications in textiles

6

Biomimetic principles in clothing technology

V. KAPSALI and P. DUNAMORE
London College of Fashion, UK

Abstract: Some key problems facing the clothing sector are explored in terms of the possibilities emerging from the principles of biomimetics. An overview is presented of the clothing sector including design process and product requirements and a brief description of biomimetic design and development methods is given. Opportunities, key issues and future trends for biomimetic innovation in the clothing industry are explored. Biomimetics can help improve the ecological footprint of the sector while inspiring innovative design through clever use of materials and structures.. The framework for biomimetic innovation is focused on the functional aspect of clothing and not on the aesthetic.

Key words: biomimetics, clothing design, clothing technology.

6.1 Introduction

Bionics, biomimesis, biomimicry, biognosis and *biomimetics* are all synonyms used in various parts of the world to describe developments inspired by the functional aspects of biological structures. However, the notion of biomimetics is often confused with *biomorphic* which refers to the visual abstraction of organic or biological forms. This was a term used to describe the work of surrealist artists in the 1930s such as Barbara Hepworth and Joan Miró. Biomimetic developments have no intentional or direct impact on the aesthetic aspects of design (Hollington, 2007); instead developments are informed by biological structures that aim to transfer functional properties of biological 'mechanisms' into the man-made world.

Biology has always been a rich source of visual and aesthetic inspiration for the design of clothing, common to every culture and era. There are countless examples of motifs such as flowers, insects and various animals, incorporated into the design of textiles either through structural patterning, print or embroidery; London designer Mathew Williamson uses floral motifs extensively in elaborate designs. Patterns inspired by plants and animals are equally as common, replication of animal markings such as the 'leopard print' has become a trademark for Italian fashion house Dolce and Gabanna.

The natural environment has provided essential raw materials for the construction of clothing for thousands of years. The incorporation of animal skins into clothing ensured human survival in conditions of extreme cold. Inuit hunters for example, were able to transfer the protective functionalities of natural skins into their clothing systems, by using combinations of seal and bird skins they create highly insulating and water resistant clothing (Ammitzboll *et al.*, 1991). Plants and animals were also the sole source of fibres (cotton, flax, silk, wool, etc.) used for the construction of textiles until the dawn of the man-made fibre industry in the early twentieth century.

Innovations inspired by a biological mechanism have rarely found application in the clothing sector. In fact, the first development linked to biomimetics was the invention of Velcro in 1950. The hooks by which burrs attach themselves to animal fur inspired this dry adhesive tape. Today, the product has many applications in industry with a strong presence in the clothing sector.

The last decade though, has seen a gradual migration of biomimetic technologies into clothing mainly through functional and performance textiles. These developments have both introduced novel functionalities to clothing, such as performance enhancement offered by Speedo's FastSkin swim suits, as well as alternative methods for incorporating additional functionality to textile systems. The Lotus Effect, for example, adds stain resistant properties to textiles that are more environmentally sound than conventional coatings and finishes (Slater, 2003).

With only a few commercially available examples, biomimetic innovation in the clothing sector is undoubtedly in its infancy. To date, there have been no publications dedicated to the investigation of the possibilities biological paradigms could offer the clothing industry from a functional perspective. In this chapter, the aim is to commence such an exploration by pairing some key problems facing the clothing sector with suggestions emerging from the principles of biomimetics. An overview of the clothing sector including design process and product requirements will be followed by a brief explanation of the principles of biomimetic design and development methods. Finally, opportunities, key issues and future trends for biomimetic innovation in the clothing industry are explored.

6.2 The technology of clothing

The relationship between the clothing industry and biomimetics requires an outline of each field before the links between the two can be identified. This section will focus on the clothing industry and will provide a brief overview of the processes involved in production of clothing from design to point of sale. In addition, the generic requirements of clothing functionality from the perspective of product performance are also described and will include clothing designed for casual use and protective apparel. This section highlights key events and concepts necessary for the purpose of this chapter; it is by no means exhaustive.

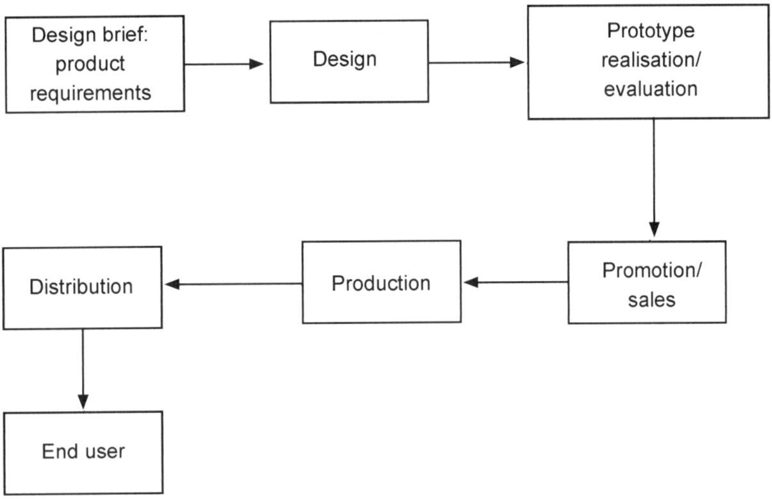

6.1 Apparel production process (source: V. Kapsali).

6.2.1 Overview of clothing design and production

The clothing sector produces a wide variety of products whose end use can range from aviation and space travel to fast fashion items that are designed to have value for a limited period of time. Various models specific to each sector have been developed to illustrate the production stages. Figure 6.1 is a simplified model highlighting the core procedure from the design table to point of sale. The field is so large that for the purposes of this study, the map illustrated in Fig. 6.1 deliberately omits the production of components, textiles, packaging and other products associated with making, storing, transporting and distribution of garments. The focus is placed on the sequence of events that occur as a result of clothing design decisions; assuming that the components, textiles, etc. chosen by the design team are based on information such as the composition, performance and history of the particular item.

The planning stage is essential to the overall product development process because it is at this point that the design brief is identified. The definition of the brief is influenced by factors such as past sales, trend predictions, customer requirements, consumer profile and functional requirements. The fashion sector places emphasis on current trends, culture and aesthetic values (Black *et al.*, 2005). Clothing whose purpose is to protect the wearer commands an advanced level of performance and technical expertise; in this case, the functional requirements compiled are essential to the design brief. This initial step in the product development process is crucial as the expectations of the product are defined; the design brief will govern the majority of decisions made during the design process.

The next stage is to realise the design brief and interpret it into a collection of garments, this may involve a single designer or an entire team. High levels of creative processes are required to interpret the brief into product ranges; this often involves a dynamic flux of idea generation and decision making. Other factors such as cost, ergonomic considerations and aesthetics inform decisions. Designers are specialised individuals who are able to decide on the optimal combinations of form and shape with components of the garment range to suit the brief requirements as well as to visualise and communicate every aspect of their ideas to production and sampling teams.

Following the design process, teams of pattern cutters, cutters and sample machinist, produce samples of the designs. This step occurs in close collaboration with the designers to ensure that the interpretation of each garment is true to the original design. The sample garments are then subjected to various tests that determine their performance and the compatibility of their components. The level of assessment varies with end use and sophistication of functional requirements; protective clothing is subjected to more stringent tests than that of 'disposable' or 'fast fashion'. This is usually the point at which the involvement of the design team ends and the work on the next project or range commences.

The time allocated for the completion of a cycle from design table to consumer varies according to the market sector; high-street chains can produce new ranges every six weeks whereas couture fashion houses produced a range twice a year. The promotion and sales period is also unique to each clothing sector. New lines are often displayed at trade exhibitions, catwalk shows such as the biannual Fashion Week that occurs in key international fashion centres such as New York, Paris, London and Milan.

The nature of buying and selling in apparel varies greatly; the client can order products from an in-house sales team or visit a sales agent that can represent a variety of product lines from different companies. Many large chains have buyers working with the design team to select the items that will be produced for retail. Again, production lead times vary greatly depending on the nature of the business.

6.2.2 Clothing system requirements

The clothing industry caters for a wide range of end users from disposable fashion to high-tech protective clothing. In order to achieve this variety, each sector uses specialist technology developed and adapted to suit each particular product specification. At one end of the spectrum, rapid response systems have been developed to produce clothing to cater for fast fashion consumers used by retailers such as Zara. At the other, garments can be engineered specifically to meet the needs of an individual such as couture garments or made to measure tailored suits. Black *et al.* (2005) developed a generic map (Fig. 6.2) to reflect the core requirements of a system of clothing common to each sector. This section will offer a brief description of a clothing system and highlight the key requirements common to conventional clothing.

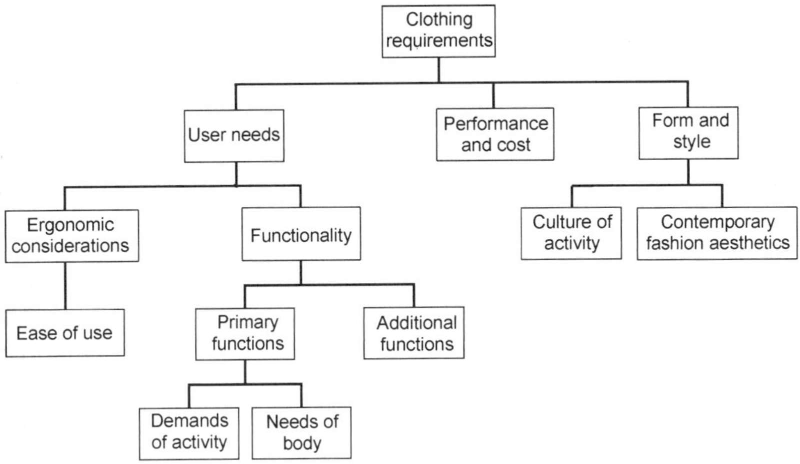

6.2 Clothing requirements (source: Black *et al.*, 2005).

A system of clothing is a very complex structure that creates a *portable environment* (Watkins, 1995) encapsulating the wearer. The role of this environment is to provide the necessary physiological and psychological conditions for an individual to operate and engage within the surrounding physical and social environment. The system can be made up of one or many layers of clothing and constitutes both the fibrous material and the air that extends from the surface of the skin to the outer face of the external garments. The micro-climate created within the system is dynamic and influenced both by external factors (external climate, activity of the wearer) and internal factors (fibre properties, textile structure, design of garment). Depending on the end use of the clothing system, emphasis on the design may be placed to cater either for psychological functions (fashion) or for physiological (performance or protective garments), however all clothing must satisfy some basic requirements.

Successful design is heavily dependent on the satisfaction of such core requirements (Fig. 6.2). The form and style of each item within a clothing system must suit the culture of the activity and meet basic contemporary design aesthetics. Although this may appear more important to the fashion sector, Black *et al.* (2005) identified several cases where individuals working in hazardous environments rejected their protective clothing because they were deemed unsuitable in terms of look and design. Evidently, the aesthetic qualities of clothing are a requirement important to all sectors and end uses necessary to satisfy the basic psychological functionality of clothing.

A clothing product needs to balance performance with cost; a successful design convinces the consumer that its price is suitable to the performance of the garment. Clothing must also satisfy basic ergonomic considerations to avoid inhibiting general life activities and functions. It is vital that a clothing system is easy to use (adding and removing garments) and does not restrict movement. Any additional functionality is determined by the design brief and the expectations of the product's end use.

The primary functions of a clothing system ensure the physiological needs of the wearer are met and the properties of the system are compatible with the demands of the activity and environment. Fire fighters require their clothing to protect them from flames yet prevent them from overheating in the extreme conditions of a burning environment and the energy produced during activity. Cold water diving suits need to sustain the core temperature of the wearer to prevent hypothermia. Urban dwellers in cities such as London or New York require insulating coats and jackets that protect them from short periods of extreme cold experienced during the winter months. These items of clothing also need to be light and durable and easily removed when they are within an enclosed public transport system, place of work or residence.

Additional functional requirements represent possible future demands from clothing enabled by new and emerging technologies. The advancing fields of *bio, nano-, electro-* textiles are introducing new properties to apparel that could supplement the functionality of conventional clothing to meet changing needs of the consumer's lifestyle. Remote connectivity, for instance, enabled by innovations in wearable electronics offers clothing that can take on additional roles currently performed by devices such as mobile phones, PDAs and satellite tracking devices. This sector of the clothing industry is very new and the first innovations that have begun to appear on the mass market have been mainly pioneered by the sportswear and performance sectors.

6.3 Overview of biomimetic design and development

The overview of the clothing industry and product requirements described in Section 6.2 provides one half of the story in the possible relationship between biology and the design of clothing. This section will highlight the key principles of biomimetic design and the methods used to develop ideas into new technology thus providing the missing component necessary to embark on the exploration of the links between clothing and inspiration from biology.

6.3.1 The principles of biomimetic design

Our knowledge of evolution depicts the natural environment as a testing ground for design development in nature. Selective pressures are exerted onto the organisms of an ecosystem, for example, through limited reserve of nutrients vital to sustain life. In order to survive, plants and animals evolve mechanisms and structures that enable them to make optimal use of minimal resources (Beukers and

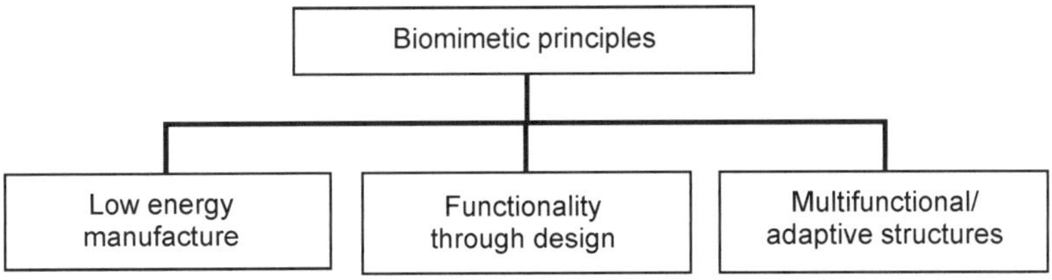

6.3 Biomimetic principles (source: V. Kapsali).

Hinte, 1998; Vincent *et al.*, 2006), thus successful 'design' survives and bad 'design' disappears.

Originally, the conceptual link between energy or resource in nature and cost in the engineering world established a bridge between the two fields. Optimal use of minimal resources was interpreted as an opportunity to develop clever yet cheap materials and structures (Beukers and Hinte, 1998). This notion of energy = money evolved to encompass the greater cost to the environment and the consumption of natural resources in the construction of man-made products (Benyus, 1997). Biomimetic scientists believe nature is a rich source of clever and sustainable design whose basic principles are illustrated in Fig. 6.3.

Functionality through design

The functional properties of biological materials are engineered through design and distribution of their basic building blocks (Benyus, 1997; Beukers and Hinte, 1998). Conventional engineering relies on the properties of the materials to deliver desired function such as stiffness, strength or elasticity to structures. Usually, whenever a new property is required, a new material is synthesised; as a result there are over 300 man-made polymers currently available. There are only two polymers in the natural world – protein and polysaccharide, whose structural variations offer a range of properties that surpass those of their man-made counterparts (Vincent *et al.*, 2006). Insect cuticle, for instance is made from chitin and protein and can demonstrate a host of mechanical properties, it can be stiff or flexible, opaque or translucent, depending on variations in the assembly of the polymer (Vincent, 1982).

Conditions of manufacture

The production of conventional man-made materials and structures generally requires great amounts of energy, high temperatures, high pressures and toxic chemicals. The man-made fibre industry is a prime example of 'high-energy', 'high-waste' production processes. Natural materials, however, are manufactured in 'low-energy' conditions employing normal temperatures and pressures no different from those necessary for life. There is no need for esoteric chemicals; usually water is adequate for the creation and growth of structures (Benyus, 1997).

Multifunctional/adaptive structures

The multifunctional profile of biological materials ensures their efficient and clever use of available resources. Structural design features are a key tool for introducing additional properties in a material. The texture of a surface for instance can evolve to provide water resistant properties to a plant or animal. Functional surfaces often occur on the upper surface of a leaf, where they protect the plant from contamination; the lotus is a plant well known for this property. Several species of insects employ functional surface textures that render their wings hydrophobic. Also, the shell of the dung beetle owes its anti-adhesion and anti-wear properties to its surface texture (Nagaraja and Yao, 2007).

In the man-made world, multifunctional materials are often created using composite technology – the property of one material is added to another to create a material with two or more properties – this is often the case in textile technology. The commercialisation of breathable, wind and water resistant membranes such as *Goretex* and *Sympatex* brands has lead to the introduction of many 'composite textiles' to the clothing sector. Layers of fabric and membrane are laminated together to create multifunctional systems, whose properties amount to the sum of the individual properties of each component, for example Belgian-based Concordia Textiles produce a composite range branded *Omniclima* which is made of three layers (a hydrophilic polyurethane film, a foam and a textile) to create systems that are breathable, water and wind resistant. Also Italian mills, Eurojersey and Lanifico Bengali create textiles using combinations of fleece and adaptive membranes as part of their *Sensitive* range.

Adaptive

There is a fundamental difference between the nature of materials in biology and those of the man-made world. Biological materials are created by the organism whereas their man-made counterparts are fabricated by external efforts (Hollington, 2007). The structure of biological materials is defined partly by DNA and partly by the environment – 'Nature and Nurture'. The survival of plants and animals depends on the ability of their structures to adapt to the changing demands of the environment; some key properties are self-assembly, reproduction, self-repair and redistribution of vital resources (Beukers and Hinte, 1998). The design brief for man-made structures, however, is predetermined and aimed to satisfy a specific set of requirements that remain unaltered during the useful life of a product.

6.3.2 Development models of biomimetic technology

Biomimetic technology is a relatively new interdisciplinary field with a short history; as a result there is currently no standard methodological approach to the interpretation of ideas. There is, however, increasing debate in the area with key

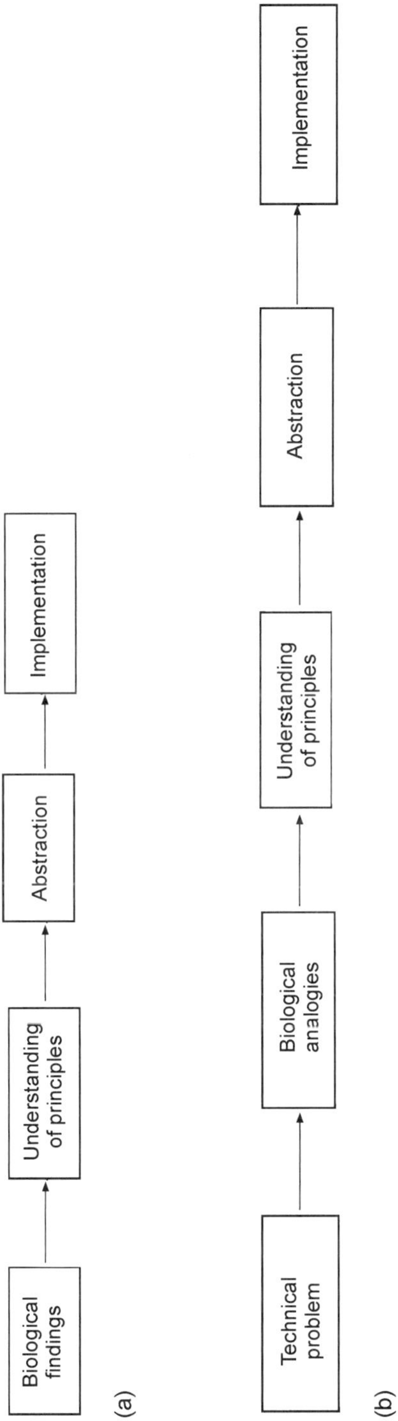

6.4 Biomimetic development model: (a) bottom up (b) top down (source: Gester, 2007).

institutions around the world developing potential models. A popular model is one adopted by the Biomimetic Guild, which illustrates a linear progression of ideas from biology to engineering (Gester, 2007). According to this method biomimetic developments can follow one of two directions: bottom up (Fig. 6.4a) or top down (Fig. 6.4b).

The bottom-up approach (Fig. 6.4a) denotes a development or innovation that has been instigated by a single biologist or a team. The biologist(s) identifies an interesting mechanism in nature that they believe would potentially have a beneficial application in industry. An understanding of the operational aspects of the mechanism is then generated in terms of functional morphology, biomechanics and anatomy. The principles are abstracted into a model that is then taken up by a team of engineers who identify methods of interpreting it into a man-made product. The Lotus Effect for instance is believed to be a bottom-up innovation originally embraced by the paint industry to create self-cleaning paint (Gester, 2007).

The top-down process (Fig. 6.4b) is initiated by industry need or a gap identified in the market. This need or gap is defined in terms of a technical problem for which analogies are sought in biology. Once suitable paradigms are identified a process similar to the bottom-up approach is followed, where a team of biologists study the mechanism(s), identify how they work and pass on the information to engineers, who interpret the ideas into solutions to the technical problem.

The model succeeds in creating a simple illustration that is reflective of the technology transfer process among biomimetic teams today. However, the model is limited by the fact that both bottom-up and top-down directions rely on a serendipitous and non-systematic approach to problem solving (Vincent and Mann, 2002). An alternative model, currently under development at the University of Bath's Centre for Biomimetic and Natural Technologies, has adopted the TRIZ (Russian acronym for Theory of Inventive Problem Solving) framework, a verified methodology used by engineers for decades that offers a systematic approach to the definition and solution of problems. Researchers at Bath University are working on adaptations to the tool, which in its traditional form is based on a database of design solutions derived from thousands of engineering patents, to include design solutions from biology (Vincent *et al.*, 2006).

6.4 Biomimetic principles and the clothing industry

The concept of biologically inspired innovation is not entirely novel to the textile and clothing industry. There have been two key occasions where attempts to imitate the properties of natural materials have led to great milestones in the history of textile technology. The properties of the silk fibre had been the object of man's obsession for centuries. Efforts to synthesise a material that imitates the strength, fineness and lustre of silk, date as far back as 3000 BC in China; it was not, however, till the early twentieth century that these efforts were successful and the first man-made fibre was mass produced: rayon imitated the lustre of silk but lacked its strength (Cook, 1984). A few decades later, the mass production of nylon caused an unprecedented revolution in the clothing and textile industry as it provided a more successful alternative to silk that was both fine and incredibly strong (Handley, 1999).

The second wave occurred in the 1970s when garments made from synthetic fibres began to fall from favour and demand for clothing made from natural materials increased. The advantages of products made from the 'new' materials such as ultra-fine cheap stockings and quick drying, no-need-to iron clothing were overpowered by the unforeseen drawbacks of the technology. Consumers increasingly complained about discomfort sensations the garments caused during wear (Kemp, 1971) such as clamminess, static and cling. The overpowering negative sensations were attributed to the hydrophobic nature of the fibres (Slater, 1977) and as a result, great efforts were made to identify ways in which synthetic fibres and textiles could imitate the comfort properties of natural fibres.

6.4.1 Biomimetic technology in clothing today

Stain-repellent textiles

The Lotus Effect™ was inspired by the protective mechanism used by the lotus leaf to prevent its surface from contamination and is, by far, the most popular biomimetic technology to have permeated the clothing sector. The technology was developed by German botanists Barthlott and Neinhuis, who discovered that the plant's self-cleaning properties were due to the surface morphology of the leaf (Barthlott and Neinhuis, 1997). This innovation was originally adopted by the paint industry to produce self-cleaning paint and later found additional application in the clothing sector in the form of stain- and soil-resistant finishes for garments.

Structural colour

Morphotex™ by Teijin Fibre Corporation is another biomimetic innovation that has recently appeared at key textile trade fairs such as Premiere Vision. Teijin researchers noticed that colour in plants and animals can be achieved using two different mechanisms, pigment and structural coloration. Pigment is the conventional method used to introduce colour to textiles and clothing but this is one of the most toxic processes in the clothing industry (Slater, 2003). Structural colour is created by the interference of reflected light caused by the morphology of the biological surface (Rossbach *et al.*, 2003). Based on the design of the wing of the South American Morpho butterfly, researchers at Teijin created a fibre made of 61 alternating nylon and polyester layers capable of producing basic colours such as blue, green and red without the use of pigments.

Performance enhancement

The drag reduction properties of shark skin were originally studied and applied to the design of aircraft skins (Vincent *et al.*, 2006); it was the competitive swimming sector though that truly adopted this technology and applied it to the design of a textile. Several swimwear companies have produced ranges using 'shark skin' fabric; the most popular is Speedo's Fastskin™ range. A single-layer clothing system employs a textile designed to imitate the functionality of shark skin to help improve the speed of competitive swimmers. Speedo claims their current version Fastskin FSII can reduce up to 4% of the passive drag during a race. However, there is no scientific evidence to support these claims and the sport's governing body FINA concluded that it was not performance enhancing. Athletes, though, claim to feel better and faster when they are wearing the swimsuit and at a disadvantage when they are not (Harding, 2004).

Adaptive ventilation

Concentration of moisture within a clothing system is a key factor in physiological comfort; conventional clothing relies on manual or behavioural methods to replenish saturated air in the microclimate of a garment. Dawson *et al.* (1997) developed a prototype textile system, based on the opening and closing mechanism of a pinecone, able to increase its permeability to air in response to an increase in microclimate relative humidity. A similar concept was recently implemented into a clothing system by Nike under the trade name Macro React[1], the technology was incorporated into a tennis dress worn by Maria Sharapova at the 2006 US Open. The garment featured a fish scale pattern on the back panel that opened as the athlete perspired to increase local ventilation and maintain the wearer's comfort.

6.4.2 Nature as a model for sustainability in the clothing sector

Increasing consumer awareness of the environmental and social cost attached to clothing has placed pressure on the industry to seek alternative methods of production as well as take on responsibility for the post-consumer stage of its products. The introduction of certification, standards and new legislation coupled with various initiatives have begun to map out new routes for improvement or indeed reform of various aspects of the industry; there is, for instance, a growing market for organic textile products. In the context of the clothing sector, organic certification ensures that the fibres (mainly cotton) have been grown without the use of toxic pesticides. Textiles made from biodegradable polymers have recently been made available to clothing manufacturers as an alternative to synthetic fibres. These are only a couple of many examples in which the industry is trying to reduce

[1]United States Patent Application 20050208860

its impact on the environment. This section will look more closely at some of the key stages of a garment's life cycle and identify opportunities where biological paradigms may provide some good ideas.

Aspects of the clothing sector have been scrutinised by various scholars and campaigners. Slater (2003) completed a comprehensive review of the environmental impact of current production and processing methods used in the textile industry. Groups such as 'Let's Clean Up Fashion'[2] and 'War on Want'[3] have exposed the social implications of cheap clothing. A study conducted by the Institute for Manufacturing at the University of Cambridge (Allwood *et al.*, 2006) combined the individual processes involved in the production of textile products and developed a model for estimating the 'energy profile' of a particular product at each stage of its life cycle.

The energy profile of a product is measured in the quantity of mega joules (MJ) consumed by the item at the five stages of its life cycle (material, production, transportation, use and disposal). The report compared the energy profile of a cotton t-shirt with that of a viscose blouse. The findings revealed that the cotton t-shirt consumed most energy during use, mainly from washing (65 MJ) whereas the viscose blouse required most resource for the manufacture of the fibre (33 MJ). The report highlights that the energy profile of garments varies greatly depending on what they are made from and how they are used. This model will be used as a framework to identify biological paradigms that may help reduce the energy profile of garments.

Pre-consumer

Often, the most costly processes have occurred before the garment reaches the consumer. The energy required for the manufacture of the textiles and components is the main contributing factor to the energy profile of an item of clothing followed by garment assembly and transportation of goods. The textile industry uses 'high-energy' processes for the manufacture of textiles and produces a great deal of waste. Both natural and man-made fibres rely on extreme temperatures, pressure and toxic chemicals for their production and processing. The cultivation of natural fibres such as cotton requires insecticides and fertilisers which pollute the soil and water. In the case of linen, although the separation of leaf and stem fibres is less energy consuming, hand retting in ditches has been replaced with high temperatures, and chemicals which speed up the process (Slater, 2003).

Recently, low energy alternatives from biotechnology have begun to replace some of these wasteful processes with enzymes. The conventional method used for the cleaning and preparation of fibre surfaces for dying, for example, requires high temperatures and toxic chemicals; the use of enzymes enables these tasks to be done at mild temperatures and without the use of any hazardous chemicals.

[2]www.cleanupfashion.co.uk
[3]www.waronwant.org

Enzymes are also replacing traditional finishing techniques such as stone and sand washing which also require great amounts of resources.

Biological materials are formed using low energy processes and require no harmful chemicals. The idea of growing garments or textiles using a biological paradigm such as skin or plants is immersed in an aura of science fiction; however, Suzanne Lee[4] currently a research fellow at Central St Martins School of Design has been collaborating with a biologist to grow garments. The project titled BioCouture uses low energy conditions to grow bacterial cellulose which is made into garments. Although the work is at a very early and experimental stage, it indicates possibilities for alternative methods of producing textiles and indeed garments.

Biological materials rely on design and assembly on a molecular scale for functionality as opposed to the nature of the raw material in the man-made world. One example is the use of microfibril orientation of cellulose in the cell walls of plant fibres. Highly orientated microfibrils produce stiff fibres that are resistant to moisture sorption, less ordered configurations are used to produce fibres that absorb larger amounts of water and swell, combinations of these two types of material are used widely in nature to power hygroscopic shape change such as the opening and closing mechanism of the pinecone (Dawson *et al.*, 1997), whereas the potential remains relatively unexplored in the textile industry.

Hygroscopic swelling of fibres is generally considered a disadvantage in the textile sector. However, this mechanism has been used to increase the water-resistant properties of a clothing system in rainy conditions. Using a tightly woven cotton fabric in the external layer of garment, during rain, the cotton fibres swell across their width as they absorb water reducing the gaps between the yarns. In turn, this improves the water-resistant properties of the garment by delaying the penetration of water from rain into the clothing system. The orientation of molecules in man-made fibres is currently managed during the later stages of spinning where the fibre is stretched; this aligns the polymer chains increasing the fibre's strength. It is possible that biological paradigms may suggest new ways of introducing properties to textiles using adaptations to these techniques and enabling the transformation of the swelling mechanism in fibres from a disadvantage to an advantage.

The use and production of surface coloration in natural materials can offer ideas for alternative methods to the highly toxic processes used for printing and dyeing (Slater, 2003). Pigment has been used to colour fibres and cloth for thousands of years, although efforts have been underway to develop non-hazardous synthetic dyes; Morphotex, has successfully transferred the structural coloration mechanism found in the Morpho butterfly into fibre technology. This invention delivers a unique method of alternative coloration that could potentially reduce the energy consumption and pollution generated during textile dyeing.

[4]www.biocouture.co.uk

Consumer use

The study conducted by Allwood *et al.* (2006) revealed that the care and maintenance of a garment during its useful life can also increase its energy profile. The data from the profile of the cotton t-shirt revealed that frequent washing cycles necessary to keep the item clean consumed more resources than the 'high-energy' processes involved in its manufacture. If an item of clothing required less washing and ironing the energy used for its maintenance would significantly decrease. The coating of fibres or textiles with compounds made from silicone or organofluorochemicals is one method currently used by the industry to protect garments from soiling and staining, but these substances are highly toxic (Slater, 2003). An environmentally sound alternative is the Lotus Effect. Enabled by plasma technology, the treatment offers a low-energy, low-pollution alternative.

The energy profile of a garment would also be significantly reduced if the useful life of the garment was extended. In today's throwaway society, consumers are more likely to replace an item of clothing once it is damaged, than repair it. Biological materials and structures are able to self-repair or self-heal, most man-made materials do not have this ability; instead they require additional resources for their maintenance and repair. A self-healing membrane[5] has been developed based on the structure of the vine *Aristolochia machrophylla* (Knott and Schampel, 2007). Although this technology is not yet suitable for introduction into textiles or fibres, it is possible that it could be implemented in the future.

Post-consumer

In 2004, the United Kingdom produced 1.5 million tonnes of textile waste. As the demand for polyester and polyamide fibres increases (Slater, 2003; Allwood *et al.*, 2006) the proportion of waste that is non-biodegradable will grow. Although natural and regenerated man-made polymers readily decompose, they do not demonstrate the properties that make synthetics so essential to the industry. Synthetic fibres are inherently hydrophobic and thermoplastic and demonstrate high tensile strength, unlike natural and regenerated fibres that are absorbent but not as strong. It would be ideal if fibres could be engineered to behave like synthetics but decompose like cotton or wool.

Polyester and polyamide fibres are conventionally synthesised from non-renewable sources. Efforts to find alternative raw materials have delivered polylactic acid (PLA) fibres synthesised from corn starch. The first PLA fibre was introduced by Kanebo Inc. under the trade name Lactron™ and, in 2003, Cargill Dow LLC (now Nature Works LLC) launched Ingeo™. PLA demonstrates many similar properties to poly(ethyleneterephthalate) (PET), which is the most common form of polyester used in the clothing industry (Farrington *et al.*, 2005). More recently, EI du Pont de Nemours and Company has launched Apexa® fibre which is also

[5]Patent no WO2007009280

made from a biodegradable polymer. Alternative methods to reduce the vast amount of energy necessary for the production of synthetic fibres are being explored by use of biotechnology. Genetically modified micro-organisms are used to synthesise polyhydroxyalkanoates (PHAs), but they are not yet commercially available (Lee, 2006).

The post-consumer processing of clothing waste is further complicated by the fibre blends and multi-component composition of garments. The industry can recycle textile fibres and reintroduce them into clothing products, Italian textile mill Figli di Michelangelo Calamai s.r.l. for instance, has incorporated fine single jersey textiles made from recycled cotton fibres into their collection since the Autumn/Winter 2007–08 season. But quality fibres can only be reclaimed from 100% compositions. Recycled fibres of mixed and unknown origin create fabrics that are bulkier and of low aesthetic value. It is now possible to create completely biodegradable garments by using polymers such as Du Pont's Apexa® that can also be made into buttons, zips, tape, etc. Biological methods such as surface morphology or fibre orientation may offer efficient alternatives to yarn blending and composite textile systems for multifunctional garments.

6.4.3 Future requirements of clothing

> 'The suit is my shape, extended, but its mind isn't mine; it's independent.' (Banks, 1993)

The functional profile of clothing is undergoing reform driven by advances in textile technology from the military and space industries. The permeation of new functional and smart technologies into other textile sectors is gradually finding a path into mass production clothing. As these new technologies offer non-conventional roles to garments, the boundaries between clothing and body are shifting. In the medical sector, textiles used for drug delivery are integrated into the body; telemedicine uses items of clothing to monitor a patient's bodily functions remotely. It is very likely that aspects of such applications will diffuse into mainstream products.

Science fiction literature offers a glimpse into the possible future of clothing functionality: some envisage garments that fulfil all the conventional requirements yet maintain an individual existence, able to change appearance, self-repair and alter their composition. Although it is not possible to accurately predict consumer demands of future clothing, they will undoubtedly be influenced by factors such as changes in lifestyle, available technology, social and political issues and changes to the environment.

Tao (2001) uses the living cell as a metaphor for future textile functionalities and forecasts behaviours such as self-repair and adaptation. We can expect clothing of the future to host an array of new properties that may interact or integrate with the body, self maintain, reproduce and self assemble to accommodate changes in our activity and environment. Materials and structures in nature already demonstrate these functions and can indicate ways of transferring the technology into clothing. Biomimetics can operate as a platform to accommodate these future requirements and provide a new perspective in the design and assembly of clothing systems.

6.5 Key issues

As industry is increasingly interested in turning to biology for ideas and solutions to key design problems, it is apparent that there is a cultural gap between biology and industry that prevents the flow of information from one specialist area to another. The language or terminology used to describe mechanisms in industry is very different to that of biology for obvious reasons. It is, therefore, very difficult for an individual in the clothing industry, for example, to describe or even imagine a problem in biological terms; accordingly biologists rarely know anything about the language used in garment design and construction.

There are few innovations that have recently found application in the clothing sector offering novel functionalities to garments. Adaptations to aspects of textile technology such as structure, finish and fibre formation have been used as a platform to introduce biomimetic developments to clothing systems. Although the application of technology from other industries is a cost effective way to introduce biomimetics to the clothing sector; persistent immigration threatens to develop 'novelty' products with functionalities that are not relevant to clothing consumers.

Biomimetic researchers can speculate on potential applications for their developments but they lack the specialist market knowledge to identify successful opportunities, so that technology push is not met by consumer pull. It is usually the biologists who identify potential mechanisms for study that they believe would have useful applications in industry. In the case of clothing and textiles, biomimetic teams have very limited knowledge of the sector and consequently their ability to identify opportunities for biomimetic innovation is limited. It is crucial that clothing designers and manufacturers work closely with biomimetic teams to identify successful projects.

6.6 Future trends

The role of clothing systems is changing, fuelled by advances in material science and textile technology, the functional profile of garments is evolving while the boundaries between wearer and clothing are blurring. Shape memory alloys have been used to alter the shape of a garment in response to heat or electrical currents. Microcapsules filled with various substances such as essential oils and synthetic wax have been successfully incorporated into foams and fibres for use in clothing. Washable electronic circuitry has been introduced into garments through textiles that enable the system to perform a range of new functionalities. Predictions suggest that future garments will imitate the behaviour of living organisms able to adapt, self heal and reproduce.

The urgent need for the clothing and textile sector to minimise its impact on the environment has driven the industry to seek alternative methods for the most hazardous aspects of their operations. Consumers are also increasingly aware of the impact their patterns of consumption have on the health of the planet and will therefore actively seek products with low-energy profiles. Biomimetics offers numerous paradigms that could help reduce the damage the production of textiles and clothing causes the environment during their lifespan.

As technology becomes more invisible, new functionality will become more conventional and everyday products of the future will interact with users to ensure their health, comfort and enhanced lifestyle. The role of the clothing designer is evolving into that of an engineer, not only creating three-dimensional structures of cloth but of a portable environment. Biomimetic principles can help steer the use and development of new technology in clothing design toward an exiting yet sustainable future.

6.7 Conclusions

In-depth knowledge of biological structures coupled with advances in micro- and nano-technologies have enabled the systematic study and transfer of useful mechanisms from biology into the man-made world. The application of biomimetic principles to engineering in general, promises sustainable, innovative functional design that makes optimal use of available resources.

In the context of the clothing industry, biomimetics can help improve the ecological footprint of the sector while inspiring innovative design through clever use of materials and structures. The framework for biomimetic innovation is focused on the functional aspect of clothing and not on the aesthetic; designers of the future will be required to take into consideration the additional functional properties that the consumers will require from their clothing.

The main obstacles that prevent the transfer of technology from biology to industry are the conceptual and cultural barriers enforced by language. Much work needs to be done in terms of building links with industry that will enable the flow of ideas, problems and solutions and thus consolidate the link between consumer pull and technology push.

Biomimetics does not always offer the most practical solution as biology does not always share a common agenda with industry; time for instance is not an important factor in nature, but it is crucial in the man-made world. The development of suitable technology can entail a long and costly process, often small adjustments to existing technology is the most effective path. Biological models may not always be the most appropriate solutions, but it is wise to explore the possibilities (Ball, 1999).

6.8 Sources of further information and advice

Centre for Biomimetic and Natural Technologies at Bath University, http://www.bath.ac.uk/mech-eng/Biomimetics/
Biomimetics at Reading University, http://www.rdg.ac.uk/Biomim/
Online resource from the US-based Biomimicry Institute, http://www.biomimicry.net/indexbiomimicryexp.htm
German Biomimetic network, http://www.biokon.net/index.shtml
Biomaterials Network, http://www.biomat.net/
Centre for Sustainable Design, http://www.cfsd.org.uk/
Sustainable Composites Network, http://www.bc.bangor.ac.uk/suscomp/index.htm
Sustainable Design Network, http://www.sustainabledesignnet.org.uk/
The Biomimetics network for industrial sustainability, http://www.extra.rdg.ac.uk/eng/BIONIS/

6.9 Acknowledgements

Special thanks to Professor Julian Vincent (Department of Biomimetic and Natural Technologies, University of Bath), Professor Sandy Black (London College of Fashion), Barbara Toll and Jane Rickard.

6.10 References

Allwood, J. M. *et al.* (2006). Well dressed? The present and future sustainability of clothing and textiles in the United Kingdom. Cambridge, University of Cambridge.
Ammitzboll, T. *et al.* (1991). *Clothing*. London, British Museum Publications.
Ball, P. (1999). 'Shark skin and other solutions'. *Nature* **400**: 507–509.
Banks, I. (1993). 'Descendant'. *The State of Art*. London, Orbit: 60.
Barthlott, W. and C. Neinhuis (1997). 'Characterisation and distribution of water-repellent, self-cleaning plant surfaces'. *Annals of Botany* **79**: 667–677.
Benyus, J. M. (1997). *Biomimicry: innovation inspired by nature*. New York, William Morrow and Company.
Beukers, A. and E. V. Hinte (1998). Smart by Nature. *Lightness; the inevitable renaissance of minimum energy structures*. Rotterdam, 010 Publishers.
Black, S. *et al.* (2005). Fashion and function: factors affecting the design and use of protective clothing. *Textiles for protection*. R. A. Scott. Cambridge, Woodhead Publishing Limited.
Cook, J. G. (1984). *Handbook of textile fibres: man-made fibres*. Wiltshire, Merrow Publishing Co.
Dawson, C. *et al.* (1997). 'How pine cones open'. *Nature* **390**: 668.
Farrington, D. W. *et al.* (2005). Poly(lactic acid) fibres. *Biodegradable and sustainable fibres*. R. S. Blackburn. Cambridge, Woodhead Publishing: 191–220.
Gester, M. (2007). Integrating biomimetics into product development. *Biomimetics: strategies for product design inspired by nature*, DTI Global Watch Mission Report: 38–41.
Handley, S. (1999). *Nylon, the manmade fashion revolution*. London, Bloombury.
Harding, D. (2004). Olympic skimpies, what our best-undressed athletes will be wearing in Athens. *Daily Mail*. London.

Hollington, G. (2007). Biomimetics and product design. *Biomimetics: strategies for product design inspired by nature*, DTI Global Watch Mission Report: 64.

Kemp, S. (1971). *The consumer's requirements for comfort. Textiles for comfort*, New Century Hall, Manchester, Shirley Institute.

Knott, B. and J. Schampel (2007). Application of biomimetics in other industries. *Biomimetics: strategies for product design inspired by nature*, DTI Global Watch Mission Report: 21–37.

Lee, S. Y. (2006). 'Deciphering bioplastic production'. *Nature Biotechnology* **24**(10): 1227–1229.

Nagaraja, P. and D. Yao (2007). 'Rapid pattern transfer of biomimetic surface structures onto thermoplastic polymers'. *Materials Science and Engineering: C* **27**(4): 794–797.

Rossbach, V. *et al.* (2003). 'Copying and manipulating nature: innovation for textile materials'. *Fibres and Polymers* **4**(1): 8–14.

Slater, K. (1977). 'Comfort properties of textiles'. *Textile Progress* **9**(4).

Slater, K. (2003). *Environmental impact of textiles*. Cambridge, Woodhead Publishing Limited.

Tao, X. (2001). *Smart fibres, fabrics and clothing*. Cambridge, Woodhead Publishing Limited.

Vincent, J. F. V. (1982). *Structural biomaterials*. Princeton, Princeton University Press.

Vincent, J. F. V. *et al.* (2006). 'Biomimetics: its practice and theory'. *Journal of the Royal Society Interface* **3**: 471–482.

Vincent, J. F. V. and D. L. Mann (2002). 'Systematic technology transfer from biology to engineering'. *Philosophical Transactions of the Royal Society* **360**(1791): 159–173.

Watkins, S. M. (1995). *Clothing: the portable environment*. Iowa, Iowa State University Press.

7

Self-cleaning textiles using the Lotus Effect

T. STEGMAIER, V. VON ARNIM,
A. SCHERRIEBLE and H. PLANCK
Institute of Textile Technology and
Process Engineering Denkendorf (ITV), Germany

Abstract: The self-cleaning property of the lotus plant, named the 'Lotus Effect', is based on the specific properties of micro- and nanostructured ultrahydrophobic surfaces, which are always completely cleaned by rainfall: the contact area of water and dirt particles is largely minimized by the double structured surface. Methods to test superhydrophobicity and self-cleaning, which differentiate between conventional soil-repellent finished textile samples and textiles that are finished with products that impose nano-dimensional structures on the fibre surface, are described. The effect of the nanostructure on the 'Lotus Effect' is investigated and applications, including architectural textiles, and future trends are outlined.

Key words: self-cleaning textiles, Lotus Effect, nanostructure, superhydrophobicity.

7.1 Introduction: basics of self-cleaning textiles

In many applications, the use of textiles is limited due to their soiling and wetting behaviour. To overcome this limitation, textiles are improved with a variety of finishes of different product classes. Recently, products have been invented that make use of the Lotus Effect and implement self-cleaning properties to a textile surface. The successful realization of this effect leads to a significant reduction in the cleaning requirement of such surfaces.

To achieve self-cleaning properties nature uses an efficient method, which has been perfectly realized on the leaves of the lotus plant.[1] Besides this species, self-cleaning properties can be found on a variety of other biological surfaces, such as cabbage, reed and nasturtium. The main function of nanostructured superhydrophobic surfaces in nature is probably the protection against pathogenic organic contamination like bacteria or spores. These are regularly removed from the leaves by rainfall.

Although discovered already in the 1970s, Barthlott and his team in the 1990s identified the reason for the self-cleaning properties and named it the 'Lotus Effect'. It is based on the specific properties of micro- and nanostructured superhydrophobic surfaces, which are always completely cleaned by rainfall: the contact area of water and dirt particles is largely minimized by the double structured surface. This in combination with hydrophobic chemistry results in extremely high contact angles that let water drops roll off at the slightest inclination, in so doing, taking up all adherent particles and removing them, leaving behind a clean and dry surface.

On many of these surfaces even high-viscous liquids (e.g. honey) drip off. The Lotus Effect is based on a minimization of the contact area of hydrophobic surfaces by an overlapping double structure approximately 100 nm to approximately 100 µm in size.

Because of this active principle, the Lotus Effect differs from the 'soil-repelling' and 'soil-release' function. As the Lotus Effect depends only on physicochemical characteristics it is independent of the living system and can be transferred into technical systems. The first commercial products with the Lotus Effect were wall painting and roof tiles. The term 'Lotus Effect' is a registered trademark for many applications.

7.2 Learning from the Lotus Effect: superhydrophobicity and self-cleaning

7.2.1 Basis for superhydrophobicity

The wettability of a surface with water and air as the surrounding medium depends on the interfacial tensions on the border of the three phases. The relation of the tensions determines the contact angle (the angle between the surface and the tangent at the water drop lying on it). If the critical surface energy of the solid is higher than the surface tension of the liquid, the liquid drop wets the solid surface well, because the wetting leads to a condition of lower energy (Fig. 7.1a). If, however, the critical surface energy of the solid is lower than the surface tension of the liquid (Fig. 7.1b), the drop assumes a spherical shape, because the spreading of the liquid would lead to a bigger area with high surface energy, thus to a condition of higher energy. A high contact angle results.

Thus, a strongly hydrophobic surface must have a very low surface energy. The maximum attainable contact angle of smooth low-energy surfaces is approximately 120°, if the surface is composed of closely packed CF_3 groups. In this instance the critical surface energy amounts to approximately 6, 7 mN m^{-1} at 20 °C.[2]

When considering rough surfaces the following two cases (Fig. 7.2) have to be distinguished. If the surface is hydrophilic (Fig. 7.2a), roughening improves the wettability with water; the water becomes sucked into the capillaries. Roughening of a hydrophobic surface (Fig. 7.2b) further reduces the wettability with water and leads to superhydrophobicity with contact angles higher than 140°.

Surfaces with an overlapping of structure elements of approximately 100 nm to approximately 10 µm in size such as those found for example on lotus leaves, but also on other leaf surfaces, thus lead to water contact angles of over 160°.[1] On such

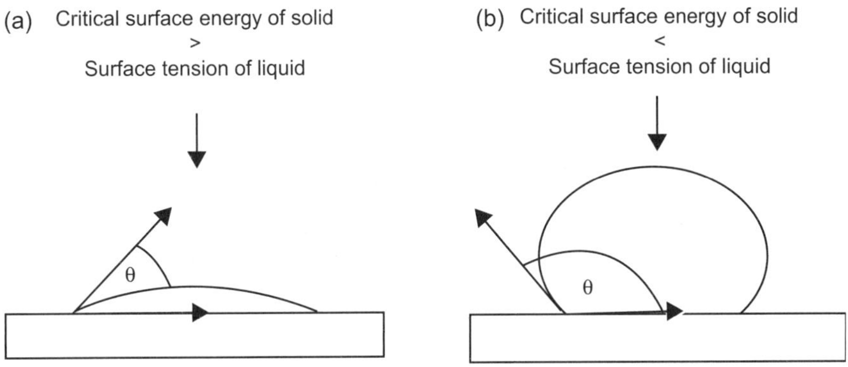

7.1 Correlation between wetting behaviour, critical surface energy and surface tension.

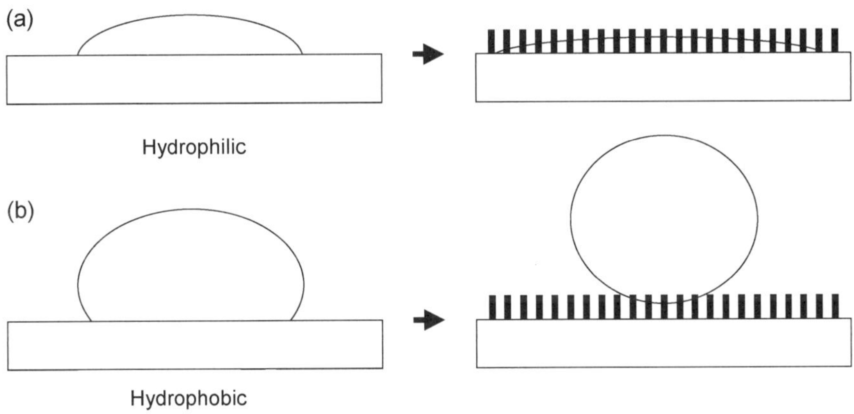

7.2 Correlation between surface roughness and wettability.

superhydrophobic surfaces, the water drop lies only on the peaks and air is trapped in the microstructures. Thus, the contact area between the solid and the water drop is minimized and the interfacial tension to air is increased. The water drop gains only very little adsorption energy to compensate for the energy which would be necessary for further wetting. This forces the water drop to adopt a spherical shape (Fig. 7.3).

Attempts have been made to describe the wetting of rough surfaces mathematically in terms of the resulting contact angles, e.g. by Wenzel[3] or Cassie and Baxter,[4] whose work is summarized in reference to. On an ideal micro- and nano rough hydrophobic textile surface, wetting with water droplets seems to be absolutely impossible (Fig. 7.4).

7.2.2 Self-cleaning function

In the technological implementation, various methods have been used to avoid the adhesion of dirt and to improve the release of dirt.

7.3 Water drop on superhydrophobic textile surface.

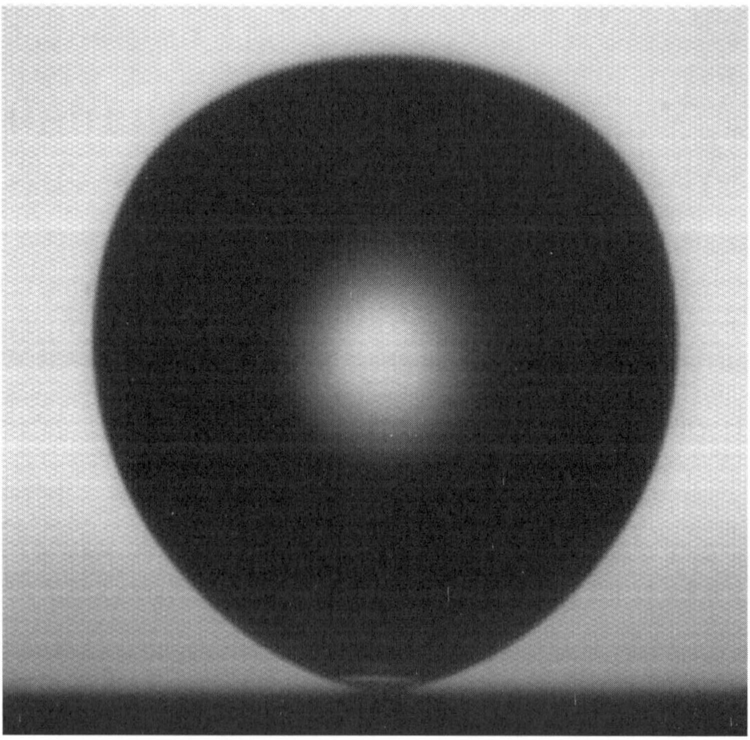

7.4 Rebounding droplet from a superhydrophobic surface.

Smooth surfaces without specific hydrophilicity or hydrophobicity

Extremely smooth surfaces show a reduced soiling behaviour because particles have only low mechanical hold and can be removed by air or liquids. However, the adhesion of residues from drying of liquids or filming cannot be prevented. To remove them, detergents (surfactants) and mechanical support are necessary. Therefore, the self-cleaning effect of smooth hydrophilic surfaces is low. With extremely smooth surfaces low soiling is sufficient to impair the aesthetic impression crucially, long before the function is limited (e.g. paints and panes).

Smooth hydrophilic surfaces

Hydrophilic coatings force aqueous drops to spread out to very thin surface films. Therefore, residues from drying up are deposited relatively evenly with little interference on the surface. Moreover, the extremely good wettability facilitates the cleaning effect of aqueous solutions. An improvement of the self-cleaning ability can be achieved by the introduction of photo-catalytic effects, as for example with the use of titanium dioxide. Organic dirt components can thereby be decomposed to low-molecular-weight products such as CO_2 and H_2O.

Smooth hydrophobic surfaces

If hydrophobic coatings are applied to smooth surfaces, the soiling behaviour can be reduced and the self-cleaning effect increased. Aqueous solutions drip off, so that the formation of residues by drying is decreased. In the textile sector, the water and oil repellency treatment of fibres with fluorine-based chemistry for the decrease of soiling of clothing and technical textiles is well examined and has reached technical maturity.

Rough hydrophobic surfaces (Lotus Effect surfaces)

If overlapping structures in dimensions of some micrometres and superposed structures of 50 to some hundred nanometres are applied to surfaces, and the chemistry of the surface is hydrophobic, soiling behaviour can be dramatically reduced and a real self-cleaning effect can be achieved. The effective surface contact area for dirt particles is extremely minimized by the surface structure and thus the adhesion is very low. Simply with drops of water rolling over the surface dirt particles are removed. Also, oily soiling can be washed off such surfaces by agitated water. When a drop of water rolls over such a surface, dirt particles lying on it are removed by adhesion to the surface of the drop.

Because of the roughness of the surface and the consequential low contact area, the adhesion energy of the particle to the solid surface is very low. Consequently, dirt particles from a superhydrophobic surface are completely removed, in contrast to a smooth hydrophobic surface where the energy is higher allowing only for the relocation of particles. For a smooth surface, the adhesion energy between particle and solid surface is relatively higher than between particle and water drop.

On rough surfaces, like textiles, the kinetic energy of a drop that falls onto it results in another positive effect for the dirt removal: particles lying in cavities of the rough solid surface are not reached by a drop that simply rolls over it. As a result of the impact when falling on the surface the drop deforms, so that it penetrates into the cavities and reaches the particles lying there.

7.3 Measuring techniques for the characteristic Lotus Effect properties

Methods to test superhydrophobicity and self-cleaning were developed to sensitively differentiate between conventional soil-repellent finished textile samples that have a smooth fibre surface on the one hand and textiles that are finished with products that impose nano-dimensional structures on the fibre surface on the other hand.

7.3.1 Water repellence test

The rate of superhydrophobicity is measured by determining the dynamic roll-off angle. In contrast to that, the static or dynamic contact angle which is used for the characterization of smooth surfaces like foils is applicable for textiles only in special cases. When dealing with micro-rough surfaces, especially where distant fibres dominate the surface structures, the contact angles can not be measured satisfactorily with optical testing methods. The dynamic roll-off angle represents the boundary value at which a liquid drop of defined volume that is placed on the inclined sample surface from a defined height rolls off the sample (Fig. 7.5). Correlations of roll-off angle and contact angles are given by Furmidge.[3]

For the test an instrument for measuring the contact angle, equipped with a tilting table and evaluation software, is used. The dynamic roll-off angle has turned out to be an important criterion, because it is readily measurable on textiles and allows a better differentiation of the samples than the static contact angle. Moreover with this method, the dynamic impact of the drop is included which brings the test closer to application.

7.3.2 Determination of the self-cleaning ability of textiles

To measure the self-cleaning ability quantitatively, a reproducible and standardized method exists at the Institute of Textile Technology and Process Engineering (ITV) Denkendorf. Self-cleaning efficiency is measured with testing methods using dirt that is known from other textile-testing standards and consists of

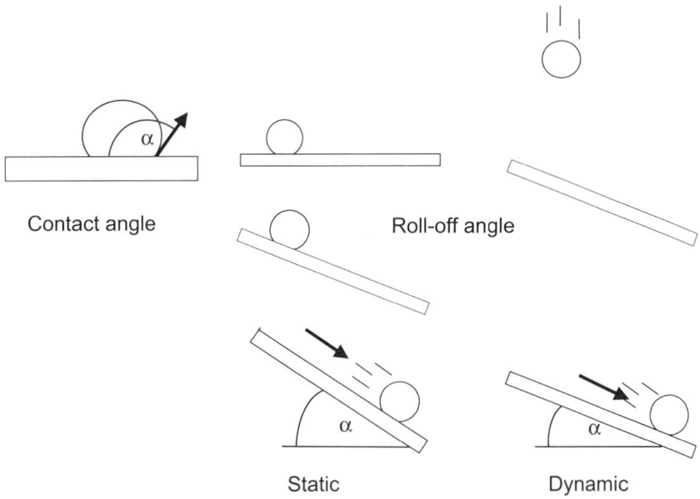

7.5 Wettability tests for the determination of Lotus Effect properties.

mixtures of different components or of a single component such as carbon black. Dirt mixtures that are used for testing include components like silica, mineral oil, olive oil and carbon black. In the applied test method the dirt is mechanically rubbed into the surface of the textiles to simulate the impact. After contamination the sample is sprayed with water.

Evaluation of remaining contamination is either done qualitatively by a stand ardized rating method using the grey scale according to the standard DIN EN 20105 A02/A03 or with quantitative methods in which the residual contaminants are detected and quantified using a microscope with image processing and subsequent particle detection software.

7.4 Technical transfer

One of the specific features of textiles in this context is that they readily bring along rough structures with at least two topological structure elements represented by the filament or fibre arrangement within the yarn structure and the yarn arrangement within the fabric structure (Fig. 7.6).

Subsequent approaches to implement the self-cleaning effect on textile-based surfaces include the optimization of fabric and yarn construction structures as well as modifying the fibre surface to low surface energies. Textile surface-finishing chemicals that meet the above mentioned requirements can for example consist of polymer-based dispersions with nano-particle additives. Other products are organic–inorganic hybrid materials based on sol–gel chemistry, some also having nano-filler additives. For successful transfer to modern textile production lines the finishing systems should be water-based. Processes for appling the chemicals

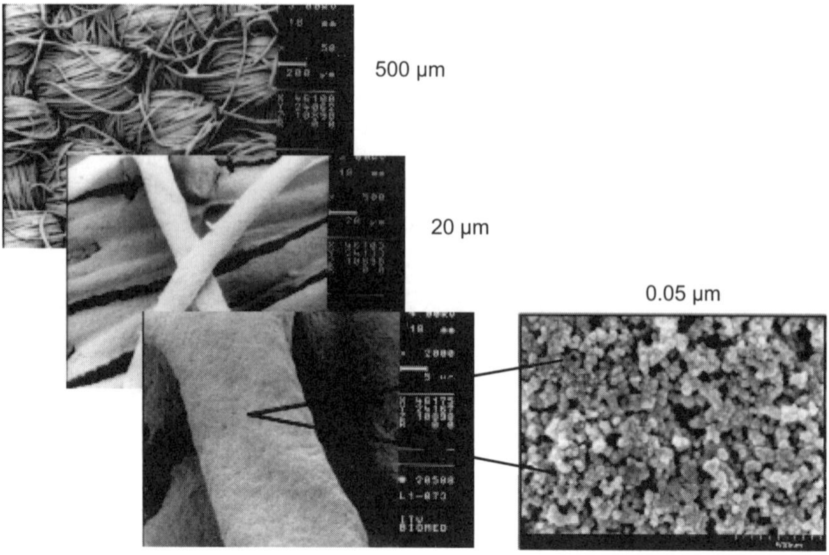

7.6 Three structure levels of textiles with Lotus Effect: fabric construction; yarn construction; fibre surface structure.

consist of standard textile finishing processes such as padding, face padding or spraying. Another attempt that is being investigated at ITV Denkendorf is to modify the fibre surface with fibre coatings at the yarn stage.[7] This process is especially interesting if finishing processes after fabric construction are limited or for sewing thread. However, dyeing processes, such as yarn dyeing or spin dyeing, have to be applied before the yarn finishing.

Textile construction also plays a crucial role in the effect of self-cleaning. In two ITV studies that were financially supported by the German Federal Ministry of Research and Technology, the influence of textile structure on superhydrophobicity and self-cleaning was investigated.

In a first study to analyze the general feasibility of the development of extremely self-cleaning textiles, the influence of construction parameters on woven fabrics made of filament yarn was investigated.[8] Particularly low wettability was measured for woven fabrics with an open yarn structure. These fabrics have distinct microstructured surfaces and show superhydrophobicity. High filament fineness supports hydrophobicity compared to yarns with thicker filaments if the yarn is constructed with low compactness so that the filaments lie side-by-side with adequate distance. Aspect rates between 1 and 2, based on distance and height differences of adjacent filaments, turn out to be especially favourable for high water repellence.

In the study to analyze the influence of structure and arrangement of staple fibres and filaments in fabrics of different types of constructions such as knitted goods, nonwovens, warp knit fabrics and woven fabrics, textile parameters, such as fibre material, yarn spinning method, filament and fibre construction, and surface

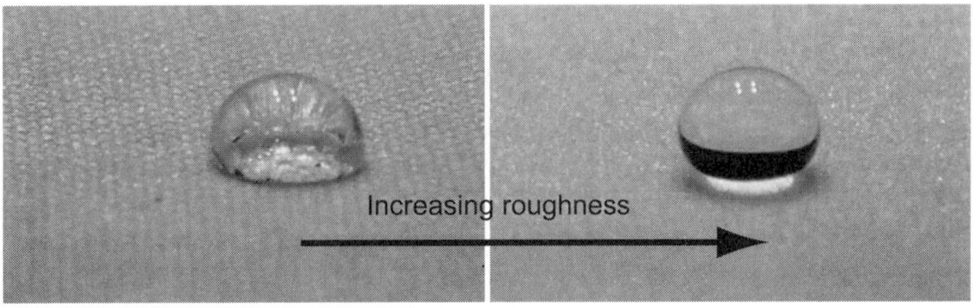

7.7 Surface wettability depends strongly on the surface roughness.

modification by mechanical and chemical methods, were varied.[9] The investigations show significant influence of hairiness of the surfaces of the samples on the water repellence. Widely spaced fibres, which are particularly present in samples made of ring-spun yarn prevent the applied small water droplets of approximately 2 mm in diameter from rolling off and therefore result in poor repellence compared to equivalent samples made of open-end yarn and Vortex-yarn. In contrast to that, the self-cleaning behaviour is not affected by widely spaced fibres and therefore the spinning method has no markable effect on it. With increased density of distant fibres the water repellence increases as contact area decreases. This can be demonstrated on flock-coated textiles (Fig. 7.7). With very dense flock fibres the water drop sits only on the fibre endings. The same effect is observed in the opposite way with singeing or calendering which results in smoother surfaces. On the other hand, for such surfaces that have less undercutting structures, the accessibility of soil particles and therefore self-cleaning ability is enhanced.

7.4.1 Influence of nanostructure

The implementation of nano-dimensional structures on the fibre surfaces enhances superhydrophobicity and the self-cleaning effect. This is shown in Fig. 7.8 and Fig. 7.9.

In this method, soot particles were deposited on the sample surface by means of a particle disperser. Then individual water drops of defined size were dropped onto the surface from a defined height. The number of particles on the treated spot were determined after each drop.

The normally smooth synthetic fibre surface can be finished with special coatings to form nanostructures that further decrease the surface contact area. Other methods to form nanostructures on the fibre are ablation processes, such as etching, embossing or the use of nano-dimensional fibres.

7.5 Applications

The implementation of self-cleaning properties on technical surfaces by using the

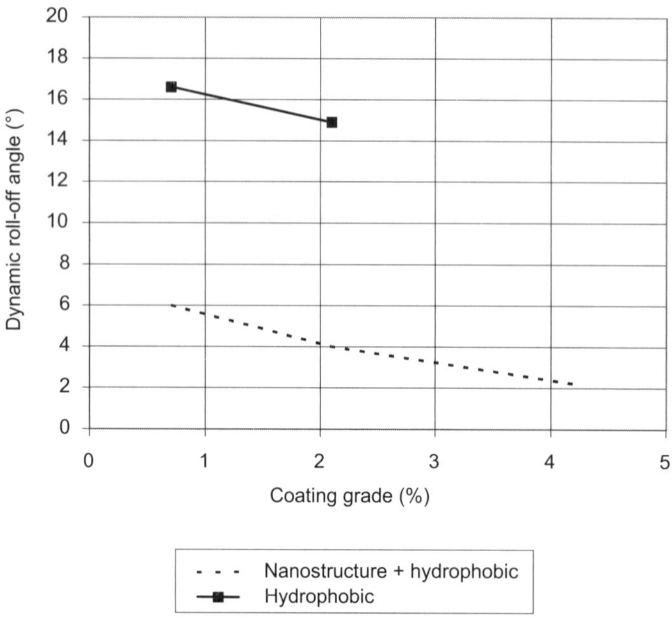

7.8 Dynamic roll-off angle; multifilament fabric with hydrophobic respectively superhydrophobic nanostructured finishing.

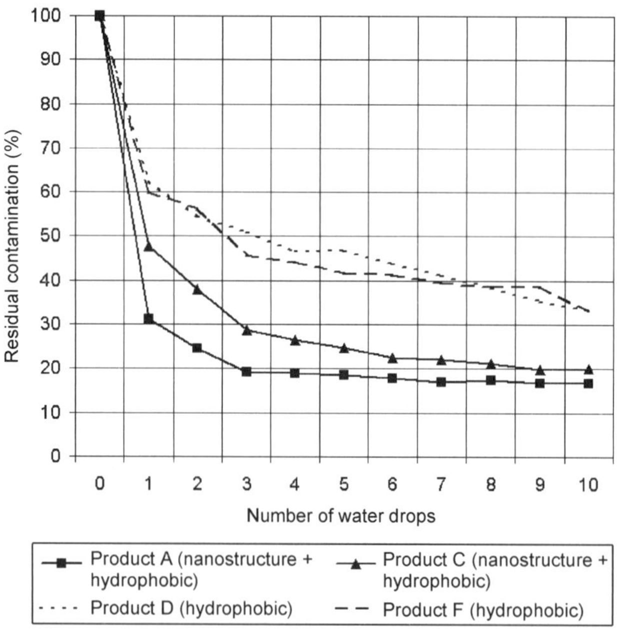

7.9 Residual contamination with soot particles after impact of water drops; multifilament fabric with hydrophobic and superhydrophobic nanostructured finishing.

Lotus Effect includes a wide potential for the development of new materials or new products and applications for known materials. For the growing market of technical textiles, a further increase in production volume, sales and application fields can be expected by a successful transfer of the Lotus Effect to textile materials. Structure-based soil- and water-repellent properties include an efficient use of materials and are therefore in agreement with the principles of sustainable development. The product lifetime is expanded and the effort required for cleaning is decreased by the self-cleaning effect.

The greatest gain lies in applications that cannot be cleaned or not without a great deal of effort, and that come into contact with rain from time to time. This includes textiles in architectural applications for weather protection, decorative or other technical functions. In many cases, the loss of aesthetic impression or other optical functions by dust, rust or the growth of algae necessitates early replacement of materials long before the loss of other technical or static/mechanical functions.

There is a variety of applications for fibre-based surfaces with self-cleaning characteristics. These include outdoor applications, such as textile roofs for airports and sports stadiums, sunscreen textiles, outdoor clothing, and indoor applications for materials that come into contact with water or water-based solutions, for example shower curtains.

Although the main requirement of the effect is the impact of rain, textiles in applications which are not exposed to the weather can be cleaned in a simple way by the impact of spraying water systems, without surfactants, and without other mechanical influences.

Whilst products for architectural uses, such as awnings with the Lotus Effect have reached market state, other applications are still in development. These are mainly textiles that are not subject to rubbing, which can destroy the surface structures that are essential for the Lotus Effect, or applications where frequent contact with water is not possible. This also applies to textiles that have to be washed frequently for hygienic reasons, for example T-shirts, shirts or bed linen.

Interesting applications in the apparel area include sport functions and safety wear. Besides the ability of self-cleaning against particle contamination in these applications, the extreme water repellence behaviour typical for Lotus Effect surfaces is the main point of interest. Examples are cycling, jogging or ski-wear.

Ongoing research and development tasks include the modification of fibres, examination and development of textile formation, surface treatment, and coatings with new nanostructured chemicals, application methods and durability tests. The durability of this superhydrophobic surface has to be measured under the conditions of product usage. Awning fabric, tested in climate-exposure test cabinets with high UV penetration for 1000 h and intermediate application-dependent mechanical load in the form of grinding exhibited better self-cleaning behaviour than awning fabric prepared with conventional finishing systems.[10]

7.10 Quality mark for self-cleaning textiles.

7.6 Future trends

To prove that textile products have a superhydrophobic and self-cleaning effect, ITV issues the Quality Mark 'selfcleaning – inspired by nature' to products that have passed the foregoing testing methods (Fig. 7.10).

Additionally, in the testing procedure for this seal of approval, the filament surface of the fibre is examined with a scanning electron microscope to analyze the surface structures, underlining the prerequisite of nano-scaled structures for the self-cleaning effect.

Recent research and development work in the field of nano-scaled surfaces on textiles at ITV Denkendorf focuses strongly on the optimization of the mechanical abrasion resistance of such surfaces. The ability for recovery of impaired surfaces, which is given for plants by growth, plays an important role in the further development of long-lasting self-cleaning technical surfaces.

7.7 Sources of further information and advice

Informative websites:
www.lotus-effect.de
www.selfcleaning.eu
www.nanopartikel.info

7.8 References

1 Barthlott, W., Neinhuis, C. Purity of sacred lotus or escape from contamination in biological surfaces. *Planta* **202**, 1–8, 1997.

2 Nishino, T., Meguro, M., Nakamae, K., Matsushita, M., Ueda, Y. The lowest surface free energy based on $-CF_3$ alignment. *Langmuir* **15**, 4321–4323, 1999.

3 Wenzel, T. N. *J* Phys Colloid Chem **53**, 1466, 1949.

4 Cassie, A. B. D. *Discuss Faraday Soc 3*, 11, 1948.

5 Palzer, S., Hiebl, C., Sommer, K. and Lechner, H. Einfluss der Rauhigkeit einer Feststoffoberfläche auf den Kontaktwinkel. *Chemie-Ingenieur Technik* **73**, 1032–1038, 2001.

6 Furmidge, C.G.L. Studies at phase interfaces. I. The sliding of liquid drops on solid surfaces and a theory for spray retention. *Journal of Colloid Science* **17**, 309–324, 1962.

7 Stegmaier, T., Abele, H., Ernst, M., Hager, T., Scherrieble, A., Schneider, P., Witt, M.-U., Wunderlich, W. and Planck, H. Functionalisation of filaments and fibres by coatings. *Technical Textiles*, **48**, March 2005.

8 Stegmaier, T., von Arnim, V. Entwicklung von Textilien mit extrem selbstreinigenden Lotus-Oberflächen. Schlussbericht der Machbarkeitsstudie im Rahmen des Förderprogramms Integrierter Umweltschutz in der Textilindustrie, Förderkennzeichen 0339935, ITV Dekendorf, 2003.

9 Stegmaier, T. Einfluss der Struktur und Anordnung von Stapelfasten sowie Filamenten auf extrem selbstreinigende Lotus-Oberflächen. Schlussbericht zum Förschungsvorhaben AIF13573N, ITV Dekendorf, 2005.

10 Keller, H., Dielemann, C. and Ebenau, A. Developments in Lotus-Effekt® finishing, Lecture on the workshop 'Nanotechnology for the textile industry', ITV Denkendorf in cooperation with Gesamttextil e.V., 19 February 2004.

8
Analysing the thermal properties of animal furs for the production of artificial furs

L. HES
Technical University of Liberec, Czech Republic

Abstract: With the continuing development and increasing production of man-made fibres, artificial furs have become a common commercial product, but they still do not act as a subtitute for animal furs in the market. This may be partly because many customers are convinced, that not only the appearance, but also the comfort properties of animal furs are better. Evidence is presented that shows that at least the thermal comfort properties of artificial furs, developed with application of biomimetic principles, can be better then those of animal furs. The thermal conductivity, thermal resistance and warm–cool feeling characteristics of 15 artificial poly(acrylonitrile)/polymer (PAN/PES) furs and 16 animal furs, differing mostly in fabric structure, were experimentally determined and the effect of hair length and diameter analysed. The warm feeling of animal furs generally decreased with the ratio of hair diameter to hair length. As expected, thermal resistance of all the furs increased linearly with their thickness. In addition, the water-vapour permeability of both kinds of fur was determined and compared. Artificial furs mostly exhibited warmer contact feeling than animal furs and offered better thermal insulation than the same thickness of animal fur. They were more permeable to water vapour than the animal furs, exhibiting a water-vapour permeability similar to that of regular cotton shirts, probably due to the open structure of the basic knitted fabric.

Key words: artificial fur, animal fur, thermal properties, biomimetics, comfort fibre.

8.1 Introduction

Properties of textile fabrics and garments include both purely mechanical properties and thermophysical (mainly heat and moisture transfer) properties. The complex effects of these properties and their interaction characterise the comfort properties of fabrics. Properties that involve the influence of fabric moisture content on selected mechanical properties, along with the effect of deformation and the skin contact force on the user's perception of garment comfort while wearing the garment, are called 'sensorial properties'. With regard to the fabric hand or handle, this is perceived by touch (tactile) and involves mechanical properties such as drape and thermal properties mentioned above that characterise the 'warm–cool feeling' of fabrics. Kawabata emphasised the importance of the latter fabric parameter.[1] Whereas the tactile properties of fabrics have been extensively studied, the 'warm–cool feeling' characteristics of fabrics and furs have been studied to a lesser extent.

Warm–cool feeling refers to our response or feeling when the skin momentarily touches a textile fabric, leather, or any material used in clothing, furniture or carpets; this is especially important if the article is in a damp condition. Since this feeling strongly affects the choices made by people when buying clothing, the objective assessment of this feeling has become very important in the last decade or so.

The main objective of this chapter is to review the thermophysiological and thermal-contact properties of 15 artificial and 16 natural furs. No similar study is to be found in the literature, despite the importance of the so-called 'biomimetic approach' for the future development of new fabrics and synthetic furs with improved comfort properties.

8.2 Brief survey of the manufacture of artificial furs

Truly artificial furs became commercially important following the availability of chemical fibres such as polyamide and poly(acrylonitrile) (PAN). Modern artificial furs exhibit several advantages over animal furs and today represent an important textile product, which in some cases cannot be substituted by animal furs.

Artificial furs can be manufactured by various procedures, such as weaving, tufting, gluing and knitting. The latest technology was developed in the USA a hundred years ago and appears to be still the most productive and economical. All the artificial furs used in the following research were knitted. This technology utilises the classical circular knitting machine, linked to a small carding machine. After drawing, the carded sliver is inserted onto knitting needles and interlaced in the knitted fabric. The resulting fur is then dimensionally fixed, mostly by coating the fabric backing or by thermal shrinking (up to 40%). The latter treatment can be achieved with steam or with dry heat, provided that the fabric contains at least 50% of shrinkable fibres. This procedure also enables the longitudinal shrinkage of certain fibrous components in the fibre cluster, in order to produce an undercoat layer in which the hairs are much shorter than the outer hairs (pile). The undercoat fibres are always much finer then the outer hairs, and as many as five different fibres are used (which are already present in the sliver), differing in composition, fineness and length, in order to provide special thermal-contact and thermophysiological properties. Thus, a very natural appearance and good thermal comfort properties can be achieved in artificial furs.

Some thermal finishing processes, such as simple or pattern thermal ironing, and unidirectional or vortex hot-air surface treatment, also facilitate the achievement of similarity with animal furs. The furs can also be naturally printed or dyed in piece.

8.2.1 Applications of artificial furs

Furs to be used for outer applications require a beautiful appearance, simulating animal fur. In this case, the hairs are longer, less textured and have greater density. Furs used as interlining are more functional, with shorter, more curled hair, and higher water-vapour permeability. Furs are also used in upholstery, shoe production, toys, the automotive industry and in medicine (for alleviating pressure sores). In this last area, the use of animal furs proved complicated or impossible.[2]

8.2.2 Advantages of artificial over animal furs

The following advantages are found for artificial compared with animal fur:

- lower price,
- lower square mass (up to 50%),
- greater durability and abrasion resistance in most cases,
- very high resistance to biological damage,
- easy dry cleaning and easy manufacture,
- higher water-vapour permeability,
- availability of any level of thermal insulation,
- has the effect of saving the lives of millions of animals (75% of animal furs are currently produced using captured animals – about 30 000 000 per year. These animals suffer from fear, cannibalism, lower immunity and killing their own young).

8.3 Experimental study

8.3.1 Device for testing thermal properties of furs

The commercial computer-controlled Alambeta device was used in this study for the fast measurement of transient and steady-state thermophysical properties (thermal-insulation and thermal-contact properties). This device works semi-automatically and calculates the statistical parameters of the measurements. The instrument also incorporates auto-diagnostics which check the precision of the measurements and avoids any faulty operation of the instrument. The whole measurement procedure involves the measurement of thermal conductivity λ (W m^{-1} K^{-1}), thermal insulation level characterised by thermal resistance R (m^2 K W^{-1}) – see equation 8.3, peak level of contact heat flow q_{max} (W m^{-2}) and sample thickness h (m). Evaluation of the results takes less than 3–5 min. The so-called 'thermal absorptivity' b (W $s^{1/2}$ m^{-2} K^{-1}) was also introduced as the objective measure of the warm–cool feeling of fabrics.[3]

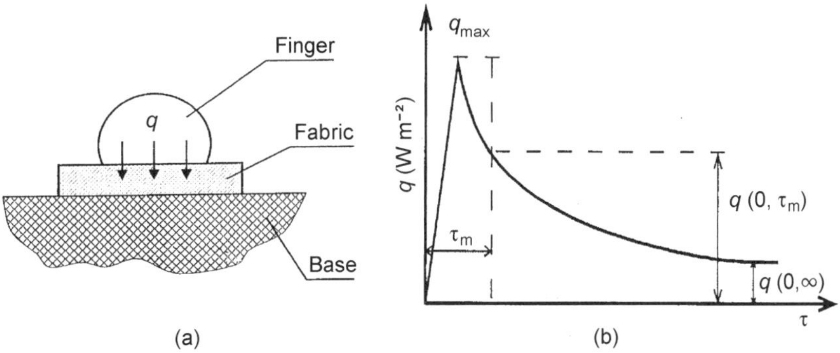

8.1 Heat flow during thermal contact of a fabric with a skin (a) and its time course (b).

8.3.2 Importance of thermal contact (warm–cool) feeling of textile fabrics

Warm–cool feeling means the feeling we get when the human skin touches any object for a short period of time, in our case textile or other fabric used in clothing, furniture or carpets. As can be seen on Fig. 8.1, when warm skin is placed on cooler fabric the heat flow between the skin and the fabric immediately increases and then slowly decreases. The peak level of this heat flow (see below) characterises this thermal contact feeling. It was found that this parameter characterises well the transient thermal feeling when we put on undergarments, shirts, gloves or other textile products. Since this feeling strongly affects people's choice when buying clothes, the objective assessment of this feeling has become very important in recent years.

Instruments for evaluation of warm–cool feeling of textile fabrics

The first device that was able to evaluate the warm–cool feeling of fabrics objectively, was developed by Yoneda and Kawabata in 1983.[1] They introduced the maximum level of the contact heat flux q (W m^{-2}) as a measure of this transient thermal characteristic, and they have published the first objectively determined values describing the thermal-contact properties of textile fabrics. Their instrument, called Thermo-Labo, was commercialised. It consists of several blocks, which are manually operated. The q_{max} parameter then depends on the composition and surface structure of the fabric, but also on the temperature gradient between the tested fabric and the pre-heated block of the Thermo-Labo instrument.

In 1986, another device for the objective evaluation of the warm–cool feeling of fabrics, but based on a different concept, was completed at the Technical University in Liberec[3] and later commercialised by the Sensora company.[4] The previously mentioned thermal absorptivity b ($W\ s^{1/2}\ m^{-2}\ K^{-1}$) was introduced as the objective measure of warm–cool feeling of fabrics.[3] The meaning of this parameter (formerly used in the civil engineering and ergonomics) is explained in next paragraph.

Thermal absorptivity in characterisation of the warm–cool feeling of fabrics

Provided that the time τ of thermal contact between human skin and a fabric is short, textile fabric is idealised to a semi-infinite body of thermal capacity ρc ($J\ m^{-3}$) and initial temperature t_2. Transient temperature field between human skin and a fabric is then given by the solution of a specific partial differential equation[3] and can be used for the calculation of the initial level of heat flow q passing between the skin (characterised by a constant temperature t_1) and textile fabric according to the next equation (the details of solution for the boundary condition of 1st order are given in reference 5):

$$q_{dyn} = b(t_1 - t_2)/(\pi\tau)^{1/2} \qquad [8.1]$$

Thus derived, thermal absorptivity b ($W\ s^{1/2}\ m^{-2}\ K^{-1}$) then follows from the relationship:

$$b = (\lambda\rho c)^{1/2} \qquad [8.2]$$

As can be seen, the level of thermal absorptivity depends neither on the temperature gradient between the fabric and skin, nor on the measurement time, but only on the contact pressure, which corresponds to the real situation. The pressure is adjustable.

The validity of thermal absorptivity as a new warm–cool feeling parameter for fabrics was confirmed by several tests where the results of relative subjective feeling of 100 people were compared with the values of thermal absorptivity found by means of the Alambeta instrument.[6] Within various research projects the thermal-insulation and thermal-contact properties of all common textile products were experimentally investigated. It was found that practical values of thermal absorptivity of dry fabrics range from 20 to 1000 (see Table 8.1). Higher values represent cooler feeling. The results show that the thermal-contact feeling of the fabrics is strongly affected by their structure and composition. It was found[7] that fibres and fibre polymers of higher moisture regain, also provide a cooler feeling. Therefore, the warmest feelings can be achieved from fabrics made from polyvinyl chloride (PVC), polypropylene (PP), PAN, whereas viscose, flax, cotton and both PAD 6 and PAD 66 fibres show the coolest feeling. Which feeling is better, depends on customer requirements: for hot summer garments, a cooler (cotton) feeling is demanded, whereas in the north of Europe warmer clothing, based on polyester (PES)/wool is preferred. As the thermal absorptivity is mainly a superficial property, its level can be changed by any superficial or finishing treatment, such as raising, brushing and coating. The instrument is so sensitive that it can reliably distinguish the warm–cool feeling of identical fabrics made of ring spun

Table 8.1 Effect of fabric structure, composition and treatment on thermal absorptivity

Thermal absorptivity, *b* ($W s^{1/2} m^{-2} K^{-1}$)	
20–40	Micro-fibre or fine PES fibre non-woven insulation webs
30–50	Low density raised PES knits, needled and thermally bonded PES light webs
40–90	Light knits from synthetic fibres (PAN) or textured filaments, raised tufted carpets
70–120	Light or rib cotton ring spun (RS) knits, raised wool/PES fabrics, brushed micro-fibre weaves
100–150	Light cotton or viscose (VS) knits, rib cotton woven fabrics
130–180	Light finished cotton knits, raised light wool woven fabrics
150–200	Plain wool or PES/wool fabrics with rough surface
180–250	Permanent press treated cotton/VS rough weaves, dense micro-fibre knits
250–350	Dry cotton shirt fabrics with resin treatment, heavy smooth wool woven fabrics
300–400	Dry VS, Lyocell, silk weaves, smooth dry resin-free heavy cotton weaves (denims)
330–500	Close to skin surface of wetted (0.5 ml of water) cotton/PP (or speciality PES) knits
450–650	Heavy cotton weaves (denims) or wetted knits from speciality PES fibres (Coolmax)
600–750	Rib knits from cotton or PES/cotton, micro-fibres knits, if superficially wetted
>750	Other woven and knitted fabrics in wet state
1600	Liquid water (evaporation effect not considered)

yarns and open end (OE) yarns. It can be also used for optimising the level of enzymic treatment. Figure 8.1 shows the heat flow with time.

An important aspect of the warm cool feeling evaluation is the change in feeling when the textile product gets wet. Because the time of the warm–cool feeling evaluation of samples in the Alambeta device is very short, less than 3 min, the evaluation of wet samples is reliable (the sample does not turn dry during the measurement). Because the thermal conductivity and thermal capacity of water is much higher than those of dry textile structures, the negative feeling of coolness of garments moistened by sweat can exceed 1000. A new field of application of this instrument is the indirect determination of the so-called 'moisture absorptivity'.[7]

Structure and properties of artificial furs

Under pressure, the fabric density increases, the orientation of the fibres changes, the thickness of furs h, and degree of porosity are reduced. Thermal resistance R becomes lower, in spite of a certain decrease of thermal conductivity (in most cases due to lower porosity) because of the relationship:

$$R = h/\lambda \qquad [8.3]$$

Table 8.2 Structure and composition of the artificial fur Afrodita

Fur structure (similar to wool 32 mm and Ambra 60 mm hair length) 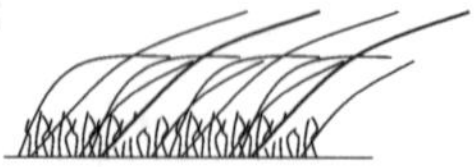	Hair length Hair fineness	28–51 mm Various fibres with fineness ranging from 3.3 to 22 dtex
Typical hair cross section	Average equivalent hair parameter	35.2 μm
Undercoat 	Hair material composition	2% polyacryl, 98% modacryl
Hairs (pile) 	Total density of knitted fabric	684 520 loops m^{-2}
	Square mass	450 g m^{-2}

Note: all the hairs are mixed in the sliver.

Table 8.3 Structure and composition of the artificial fur Daniel

Fur structure (similar to Katka with 12 mm hair length) 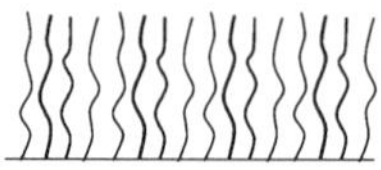	Hair length Hair fineness	20 mm From 3.4 to 17 dtex
Typical hair cross section	Average equivalent hair parameter	26.6 μm
Polyacryl 	Fur composition	15% polyacryl, 85% polyester
Polyester 	Total density of knitted fabric	652 400 loops m^{-2}
	Square mass	830 g m^{-2}

Note: backing fixed by resin.

Table 8.4 Structure and composition of the artificial fur Norsko

Fur structure (similar to Bakara 8 mm, Javor 6 mm, Brigita 8 mm hair length) 	Hair length Hair fineness	10 mm 3.4 dtex
Typical hair cross section 	Density of knitted fabric	767 496 loops m^{-2}
	Square mass	370g m^{-2}

Note: hair material composition is 100% polyacryl.

Table 8.5 Structure and composition of the artificial fur Otmar

Fur structure (similar to Upal 32 mm hair length) 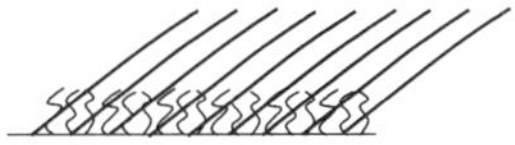	Hair length Hair fineness	35 mm 4.4–22 dtex
Typical hair cross section	Equivalent hair diameter	46.6 μm
Undercoat 	Hair material composition	40% modacryl shrinkable, 60% modacryl heat set
Outer hairs 	Density of knitted fabric Square mass	651 408 loops m^{-2} 530 g m^{-2}

Note: different length of hairs by shrinking.

Table 8.6 Structure and composition of the artificial fur Teran

Fur structure 	Hair length Hair fineness	10 mm 3.4 dtex
Typical hair cross section 	Equivalent hair diameter Hair material composition Density of knitted fabric Square mass	9.7 μm 100% polyacryl 716 284 loops m^{-2} 340 g m^{-2}

Note: thermal finishing by tamblering.

To maintain high thermal resistance, the fur hairs should exhibit a certain resistance to compression. Since this resistance increases to the 4th power of the hair diameter, even coarse fibres are sometimes used as components of the fur hair. Fine fibres and micro-fibres reduce the heat flow passage, owing to higher absorption of infrared radiation and lower heat conduction along the fibres.[8] That is why the undercoat layer of artificial furs often simulates the structure of animal furs and is made up of very fine fibres – see Tables 8.2–8.6, which show the structure and composition of furs studied, which were manufactured by the Czech BONEKA Company.

Figure 8.2 shows the effect of fur compression on the contact area. Fine fibres can easily be compressed, resulting in a large contact area. More heat is then conducted away from the hand and the contact feeling is cooler. Thus, fine fibres in a free state provide better thermal insulation than coarse fibres, but when compressed (in the case where is no system to prevent their compression), their thermal resistance decreases rapidly.

In contrast, when compressing coarse or textured fibres (hairs), the increase of the contact area is much smaller. This leads to a warmer feeling (this is discussed in more detail in the following sections).

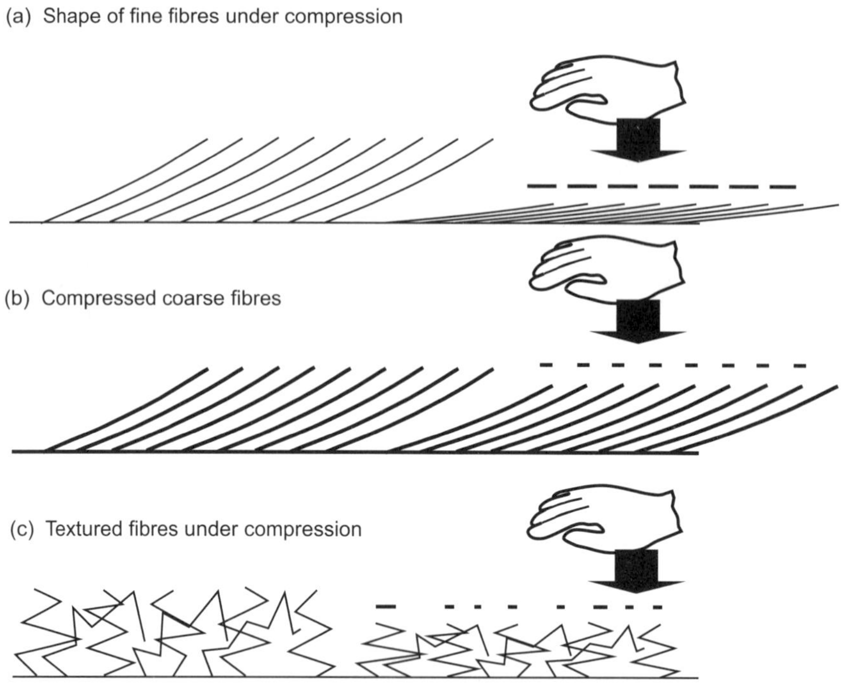

8.2 The effect of contact force on fibre deformation for (a) untextured fine fibres, (b) untextured coarse fibres and (c) textured fibres.

8.3.3 Results of thermal conductivity measurements

The thermal conductivity of various artificial furs is displayed in Fig. 8.3. Correlating these values with the structure of the furs, we can conclude that the lowest (best) levels are achieved where the fine hairs are perpendicular to the heat flow direction (see Teran fur). Moreover, the hairs should create many closed pores. In contrast, when the heat flow passes along the hair (as in the case of Daniel fur), then the thermal conductivity is higher (worse).

As can be seen in Fig. 8.4, comparing the thermal conductivity of artificial and animal furs, most artificial furs exhibit higher (worse) levels of this property. This characteristic has been explained by Vorlova.[9]

8.3.4 Thermal resistance of furs

As stated in the previous section, thermal resistance R is calculated by the ratio of thickness h and thermal conductivity λ. This thickness depends on the contact pressure. The effect of the contact pressure of the measuring head on the thermal

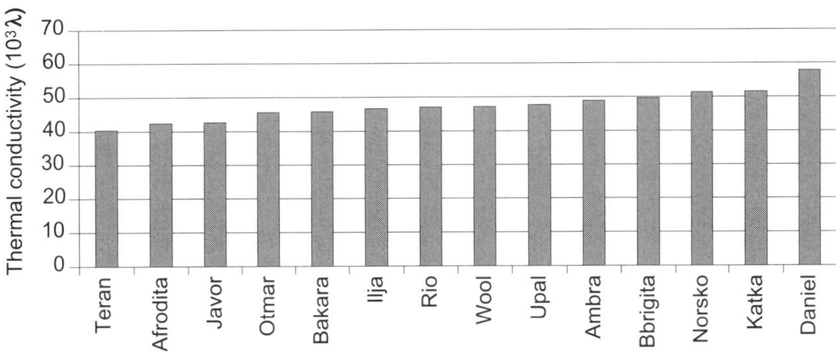

8.3 Thermal conductivity of artificial furs.

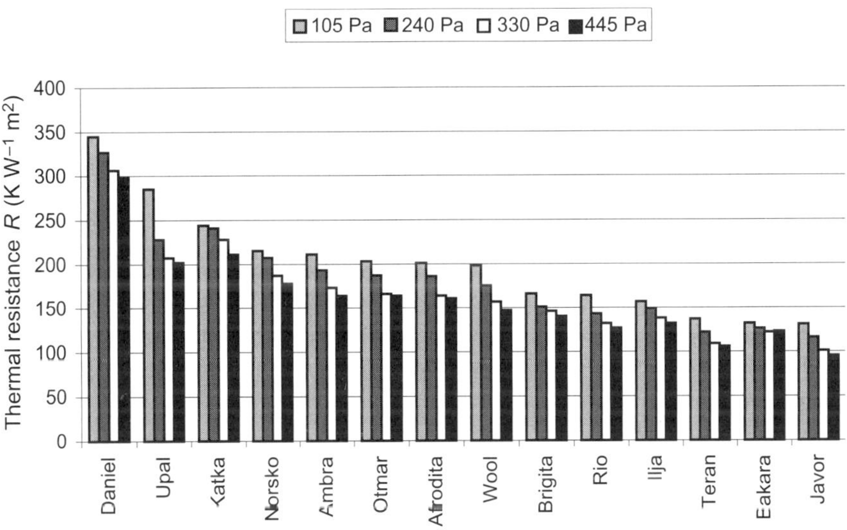

8.4 The effect of contact pressure on thermal resistance of artificial furs.

resistance of artificial furs is shown in Fig. 8.5. Data collected at four different pressures allow the thermal resistance at zero pressure to be easily extrapolated. It is clear that the technology of artificial furs enables furs to be designed with very good resistance to compression.

Comparison of the thermal resistance of artificial and animal furs, shown in Fig. 8.6, demonstrates that the thermal resistance of artificial furs is generally higher than that of animal furs, and that it can be as high as required, despite higher levels of thermal conductivity.

8.3.5 Effect of hair parameters on thermal contact area

Artificial and animal furs can be idealised on a system of cylindrical beams,

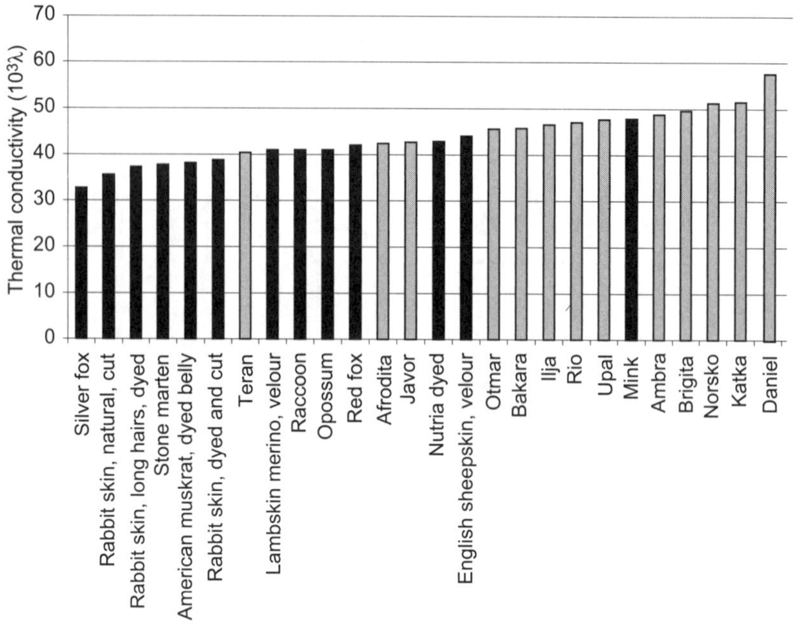

8.5 Thermal conductivity (W m^{-1} K^{-1} of artificial (shaded bars) and animal (black bars) furs.

anchored in the skin at a certain angle. Moreover, these beams are curved and their ends are frequently located almost parallel with the skin. When touching the skin, the contact force F acting more or less perpendicularly on the hair end, causes bending deformation y, which is the level of the fur compression towards the skin surface. The deformation y depends on the hair length l and hair diameter d, according to the generally recognised relationship, in which E indicates Young's modulus and J is the moment of inertia:

$$y_{\max} = y_{\mathrm{B}} = \frac{Fl^3}{3EJ} = \frac{64Fl^3}{3E\pi d^4} = Cl^3/d^4 \qquad [8.4]$$

From this relationship, it follows that the hair deformation (and also the fur compressibility) are proportional to the ratio l^3/d^4. It can be concluded that (under pressure) long and fine hairs will create a larger contact area, which results in a cooler contact feeling than in the case of short and coarse hairs.

Professional furriers differentiate the hand-feel of various furs by means of the fineness coefficient, given by the ratio d/l. As follows from the previous simple analysis, the application of such an empirical coefficient is quite well justified. Given this observation, the thermal absorptivity and conductivity of the furs analysed in this study will also be correlated with this empirical coefficient.

Table 8.7 shows that, as expected, the thermal resistance of furs increases with thickness. But this thermal resistance, calculated by the relation $R = h/\lambda$, also

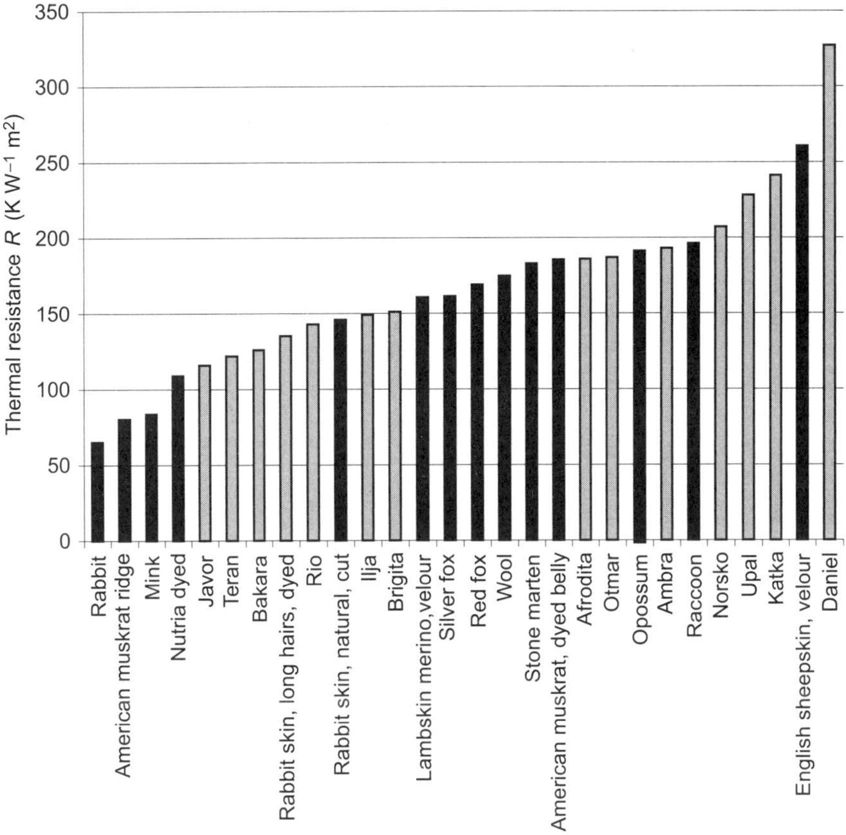

8.6 Comparison of thermal resistance of artificial (shaded bars) and animal (black bars) furs.

depends on thermal conductivity λ. When comparing the values of λ of the various furs mentioned in Table 8.6 we find that the λ of silver fox fur is incredibly low. This 'natural miracle' results from the special structure of the fur of this animal, which lives in a severe arctic climate. The hairs are hollow (confirmed by microscopic technique at TU Liberec[8]), but the cavity is separated into individual cells in order to prevent even very low heat transfer by natural convection. Deeper analysis of this fur is recommended for the future biomimetic development of artificial furs.

The study (Fig. 8.7) of the effect of hair parameters on the thermal absorptivity of furs demonstrates that the warmest feeling is achieved when the ratio d/l is 2–2.5. This result can be applied in the design of synthetic furs with the warmest contact feeling.

As indicated in Fig. 8.8, the thermal absorptivity of artificial furs varies according to the structure of the fur and fineness of the outer fibres (hairs). In this case, the effect of fibre fineness acts conversely to that of thermal conductivity

Table 8.7 Effect of hair geometry on the thermal properties of animal furs

Type of fur	Average value (CV in %)				
	Length l (mm)	Diameter d (μm)	Hair fineness coefficient d/l (μm mm^{-1})	Thermal conductivity λ (mW m^{-1} K^{-1})	Thermal absorptivity b (W s$^{1/2}$ m^{-2} K^{-1})
Silver fox	91.6	77.77	0.85	32.76(2.14)	49.55(14.11)
Red fox	70	71.44	1.02	42.06(2.78)	83.25(10.56)
Chinchilla	31	32.53	1.05	39.02(2.00)	87.83(4.09)
Opossum (North American)	90	109.49	1.22	41.04(4.17)	74.93(7.03)
Raccoon	66	93.70	1.42	41.03(1.36)	78.18(3.12)
Rabbit skin, long hairs, dyed	45.8	86.38	1.89	37.31(3.48)	62.46(8.87)
American muskrat, ridge	34	76.16	2.24	42.25(2.34)	116.10(7.04)
American muskrat, dyed belly	32	73.92	2.31	38.14(1.63)	69.38(4.09)
Stone marten	33	86.57	2.62	37.78(3.23)	69.56(12.58)
Lambskin merino, velour	12.1	33.06	2.73	40.96(1.81)	59.01(2.46)
English sheepskin, velour	15	41.64	2.78	44.06(1.36)	68.12(3.78)
American muskrat, belly	26	78.14	3.01	39.00(1.63)	107.20(4.09)
Rabbit skin, natural, cut	20.6	62.72	3.04	35.33(1.40)	69.28(4.32)
Rabbit skin, dyed and cut	16	53.38	3.34	38.79(1.03)	98.42(7.85)
Nutria dyed	42	152.21	3.62	42.88(2.15)	104.45(12.2)
Mink	19	82.27	4.33	47.87(1.19)	133.90(2.79)

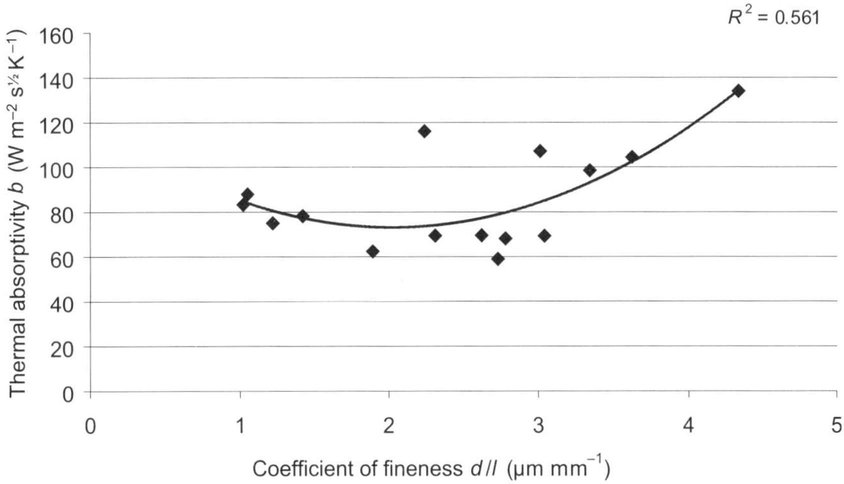

8.7 The effect of hair geometry on thermal absorptivity of animal furs.

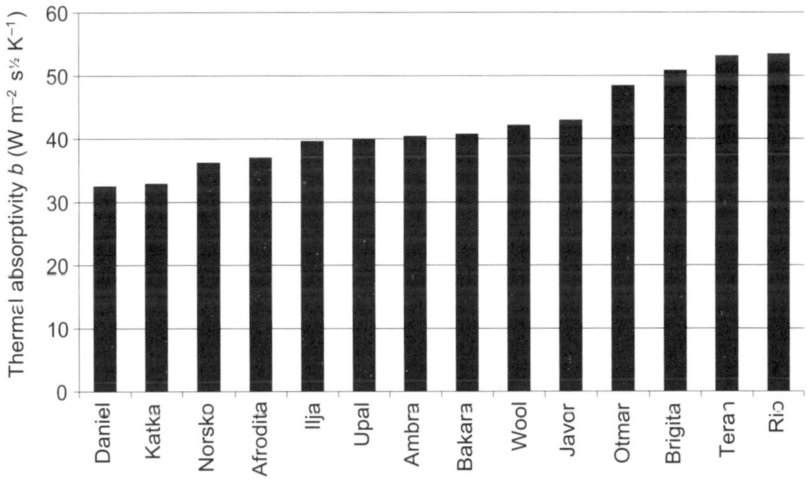

8.8 Thermal absorptivity (warm–cool feeling) of artificial furs.

requirements: hairs oriented along the heat flow direction offer a lower contact area and therefore exhibit a better (warmer) feeling than furs with hairs located perpendicularly to the heat flow direction. Fortunately, versatile artificial fur technology enables nearly any level of the warm–cool feeling to be achieved.

The levels of thermal absorptivity (warm–cool feeling) of artificial and animal furs are compared in Fig. 8.9, which demonstrates that artificial furs generally exhibit a much warmer feeling than animal furs.

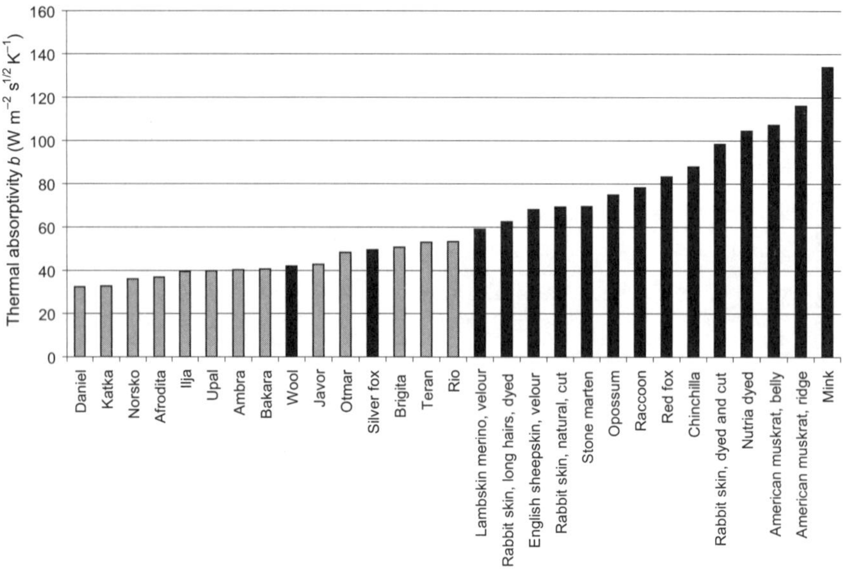

8.9 Comparison of thermal absorptivity of artificial (shaded bars) and animal (black bars) furs.

8.4 Water-vapour permeability of furs

Water-vapour permeability of fabrics represents, along with fabric thermal resistance, the most important characteristic of clothing comfort. That is why increased attention has been paid to this parameter in recent decades. Testing of this parameter by means of any common testing instrument is time consuming and requires special samples of large dimensions cut from pieces of fabrics. Unfortunately, animal samples used in this study were of small dimensions. That is why the only instrument which enabled the testing on furs was the non-destructive fast working Permetest skin model.

Results of measurement can be expressed in units defined in the ISO Standard 11092, but also in the form of the relative water-vapour permeability P, which was used in this study. The principle of this instrument is shown in Fig. 8.10. Here, the slightly curved porous surface is moistened (either continuously or on demand) and exposed in a wind channel to a parallel air flow. A tested sample is located on a semipermeable layer covering the metallic wetted area of diameter about 80 mm and characterised by high thermal conductivity. The amount of evaporation heat of liquid water taken away from the active porous surface is measured by a special integrated system. Thus, a very low time constant was achieved for the whole system, resulting in a short measurement time (full signal is registered within several minutes).

At the beginning of the measurement, the heat flow value q_0 without a sample is

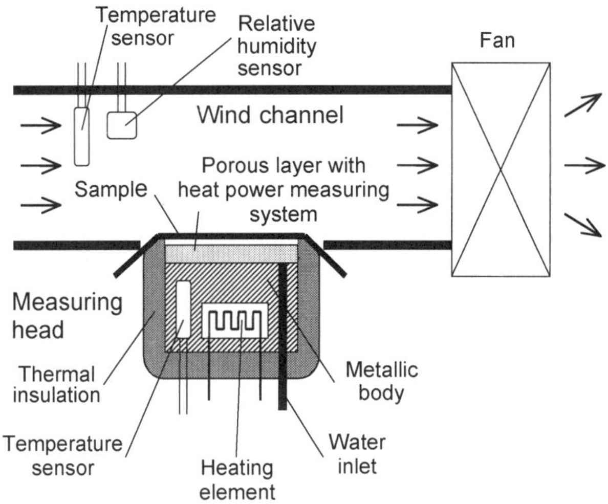

8.10 Measuring facility.

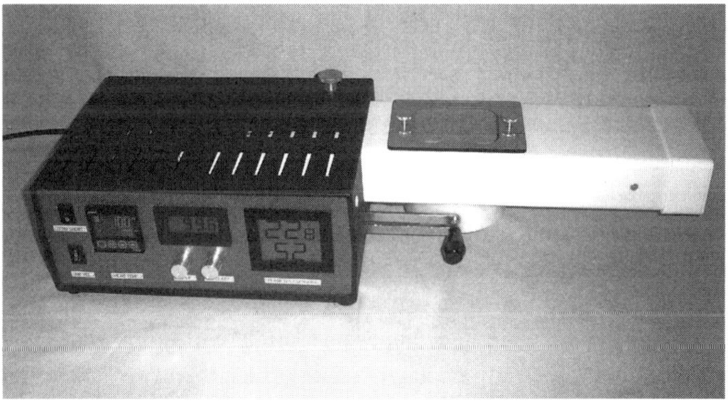

8.11 Permetest instrument.

recorded. In the next step, the measuring head is covered by the tested sample and the heat flow level q_1 is recorded. The instrument is shown in Fig. 8.11.

Relative water-vapour permeability P of the textile sample is calculated from the formula:[10]

$$P[\%] = 100\ q_0/q_1 \qquad [8.5]$$

A value of 100% permeability here represents the level of evaporation from a free water mirror, whereas the permeability relating to this reference level includes the permeability of the fabric itself, permeability of the 1.5 mm air gap and permeability of the boundary layer at walking velocity of 3.6 km h^{-1}. Given that the typical relative water-vapour permeability of new denim jeans is about 20%, then the

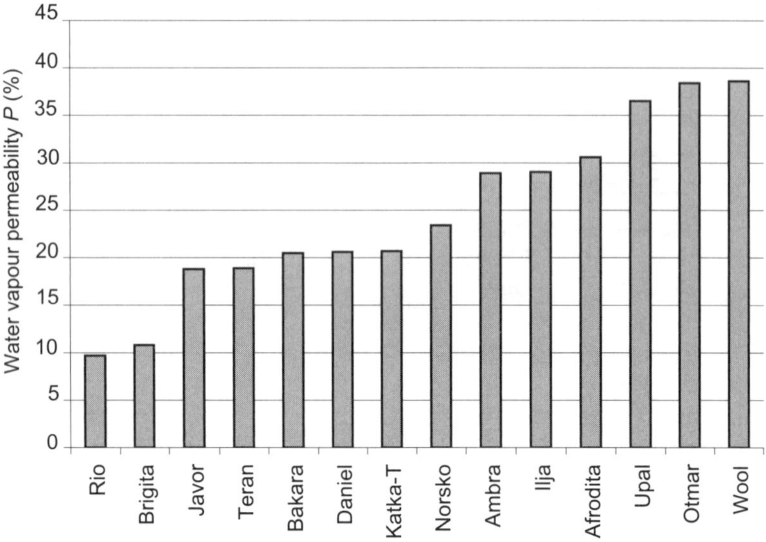

8.12 Relative water vapour permeability of artificial furs.

permeability of furs is generally high (Fig. 8.12). The lower the density of the basic knitted fabric, the higher the permeability. Compared with these artificial furs, the permeability of animal furs is quite low – 5% on average. Some furs, such as buffalo, are almost impermeable. Nevertheless, the thermophysiological properties of these furs are still very high because of their special advantages: extremely high moisture absorption, resulting in additional warmth for the wearer in cold, wet weather and cooling for the wearer on hot, dry days. Artificial furs lack this characteristic of water and water-vapour management.

Some results presented in this chaper have already been presented by Hes.[11] In this chapter, some imperfections have been corrected and additional information has been included.

8.5 Conclusions

Experimental results presented in this paper demonstrate that artificial furs generally exhibit higher thermal resistance, warmer contact feeling and higher water vapour permeability than animal furs. Their lower weight, lower price and easier maintenance present further advantages over animal furs. Nevertheless, the improved appearance and high price of animal furs have transformed them into status symbols like luxury cars. Despite this, the enormous potential of artificial fur technology and the use of new comfort fibres should open up new markets for the latest generation of artificial furs, at least in technical areas of use, such as sports, the food industry, military applications and other areas where protection against the cold is required.

8.6 References

1 Yoneda M. and Kawabata S., Analysis of transient heat conduction in textiles and its applications, Part II, *J. Text. Mach. Soc. Jpn* **31** (1983) 73–81.

2 Koryntova L., Thermal properties of artificial furs, MSc Thesis, Tech. Univ. of Liberec (2003).

3 Hes L., Thermal properties of nonwovens. In: Proc. INDEX 1987 Congress, Section B1, Geneve (1987).

4 Catalogues of the Alambeta and Permetest instruments, Sensora Co., Czech Rep.

5 Hes L. and Dolezal I., New method and equipment for measuring thermal properties of textiles, *J. Text. Mach. Soc. Jpn* **42** (1989) T124–128.

6 Hes L., Araujo M. and Djulay V., Effect of mutual bonding of textile layers on thermal insulation and thermal-contact properties of fabric assemblies, *Textile Res. J.* **66** (1996) 245–250.

7 Hes L., A new indirect method for fast evaluation of the surface moisture absorptivity of engineered garments. In: 'Internat. Conference on Engineered Textiles,' UMIST, May 20–22 (1998).

8 Hes L., Thermal resistance and thermal-contact characteristics of nonwovens. In: Nonwovens, BTRA monograph series, Bombay 1990, ed. S. M. Betrabet.

9 Vorlova L., Dimensional characteristics of furs, MSc Thesis, TU Liberec (2001).

10 Doležal I. and Hes L., P-TEST – Computerized instrument for testing of the water vapour and thermal resistance of fabrics. In: CD Proc. of 2003 IEEE Internat. Symposium on Industrial Electronics, Rio de Janeiro, Brazil, June (2003), p. 5.

11 Hes L., Thermophysiological properties of furs. In: Congress on Technical Textiles, Oct. 2004, UPC Terressa (2004).

9

The role of plant stems in providing biomimetic solutions for innovative textiles in composites

M. MILWICH and H. PLANCK
Institute of Textile Technology and Process Engineering Denkendorf (ITV), Germany

T. SPECK and O. SPECK
Universitaet Freiburg, Germany

Abstract: The significance of inspiration from nature for technical textiles and fibrous composite materials is demonstrated by examples of technical solutions that either parallel biology or are inspired by biological models. Two different types of biomimetic approach are briefly presented and discussed for the 'technical plant stem', a biomimetic product inspired by a variety of structural and functional properties found in different plants. The most important botanical role models are the stems of the giant reed (*Arundo donax*, Poaceae) and of the Dutch rush (*Equisetum hyemale*, Equisetaceae). After analysis of the structural and mechanical properties of these plants, the physical principles were deduced and abstracted and finally transferred to technical applications. Modern computer-controlled methods for producing technical textiles and for structuring the embedding matrix of compound materials render unique possibilities for transferring the complex structures found in plants into technical applications. This process is detailed for the 'technical plant stem,' a biomimetic, lightweight, fibrous composite material based on technical textiles with optimized mechanical properties and a gradient structure.

Key words: *Arundo donax*, biomimetic textiles, *Equisetum hyemale*, 'technical plant stem', smart composites.

9.1 Introduction

Modern biomimetics is a systematic approach taken by collaborating researchers from different scientific backgrounds in developing new ideas as a result of looking at an abundance of biological role models. Most helpful and catalytic for this kind of collaboration has been the development of new measuring methods and instruments, which are able to unearth natural principles or regularities hitherto unknown. The biomimetic process is very transdisciplinary, encompassing approximately seven different fields of bionic research (according to the German Federal Bionics Competence Network BIOKON): (1) architecture and design; (2) lightweight construction and materials; (3) surfaces and interfaces; (4)

fluid dynamics, swimming, and flying; (5) biomechatronics and robotics; (6) communication and sensors; and (7) optimization. The boundaries between individual fields, however, are often somewhat blurred, and the transition between subdivisions is gradual (Speck and Harder, 2006; Speck *et al.*, 2006a).

For lightweight construction and materials, very often fiber-reinforced plastics (composites) are first choice. In many instances textiles from high performance fibers are used, because they are easier to handle, need less work and are less expensive in the process of fabrication of a composite part than Prepreg materials.

Thus, in the field of composites and in many other applications of technical textiles, textiles are no longer just simply a means of producing clothing, seat covering, or carpets. So-called 'smart' textiles with high functionality conquer the market. New fiber developments are stronger, chemical- and fire-resistant, electrically conductive, or adjustably biodegradable. Using high-tech fibers, textiles nowadays are used for environmental pollution control or lifesaving implants (Milwich and Mueller, 2005). Snowboarding jackets can play MP3s and have incorporated GPS transmitters (Stollbrock, 2005). Incorporated sensors in carpets, called 'thinking' or 'smart' carpets, give emergency light, detect the heat of a fire, call a nurse if a patient falls and remains on the ground, and detect footsteps of an intruder (Lauterbach, 2005).

A common research topic for all these examples is the integration of 'smart' functionalities that preferably should be integrated into the fiber or the textile as deep as possible. That means that any 'smart' functionality should not be a rigid, sharp-angled form, which stands out from the textile or fibers as a foreign body, but should be as smooth as the textile surrounding it. Thus, the smart function itself should be made of the special fibers or textiles. In so doing, we come full circle back to nature's soft, flexible, and adaptive structures. A good example for the deep integration of a smart function is the Baby Body Clothing, developed by ITV Denkendorf (Linti and Horter, 2005). Without any hard edges or planes, it records body functions of the baby and calls for help if breathing or heart beats are unsteady (Fig. 9.1).

In the background of these examples, nature's own 'smart functions' are a broad source for textile solutions and textile applications. Biologists in Freiburg, Dresden, and Bonn, with their expert insights into special plant and animal functionalities, cooperate with ITV Denkendorf in various biomimetic projects striving for textile solutions (Stegmaier *et al.*, 2004; Harder, 2006).

With the use of strong, specially developed fibers such as glass, carbon, or aramid and their embedment in polymer or ceramic matrices, very strong and lightweight composite materials and structures can be processed. An advantage of using flexible fiber material is that they can be laid exactly in the direction of the strain lines of a designed structure. Fiber composites can be found in airplanes, space shuttles, or in racing cars. American architect Peter Testa has proposed building a skyscraper of thick strands of helically and circumferentially wound carbon fiber composites arranged similarly to a mesh, with the spaces in between the strands filled with glass plates.

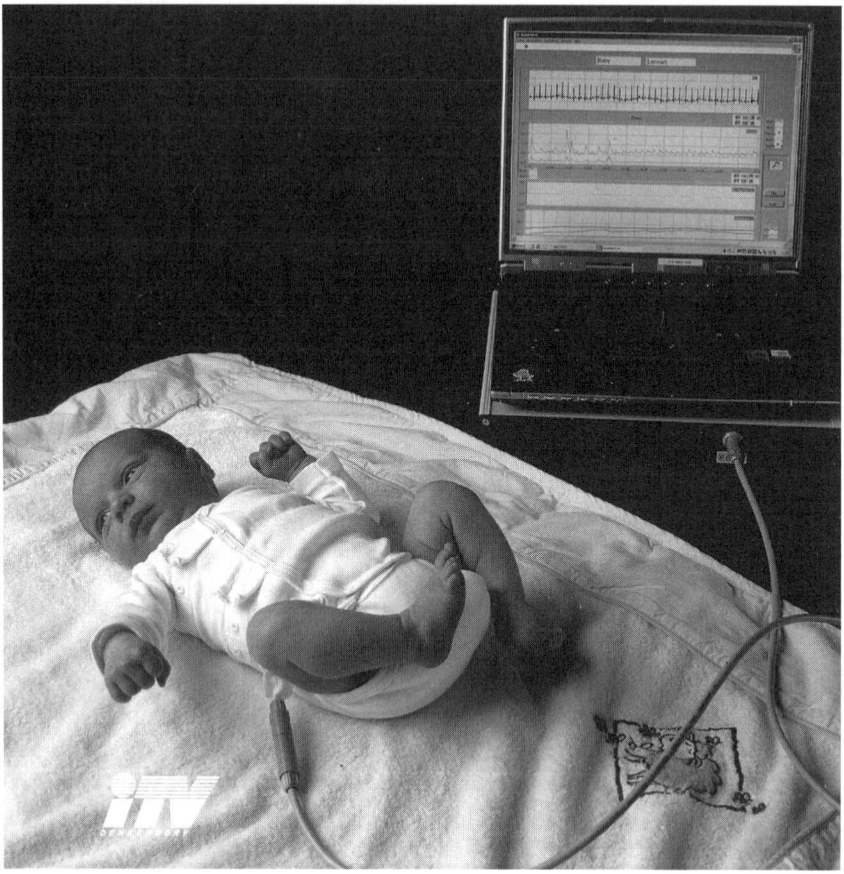

9.1 ITV Denkendorf Baby Body recording body functions.

Fiber composites can thus be regarded as functional technical textiles and are a classic example for biomimetic translation of nature's wisdom into technology. Many principles of composites have their counterpart in nature. Bones, plant stems, and wood have highly optimized the use of fibers in the exact directions of effective loads (Mattheck, 1990, 1996, 1998; Vogel, 1998) and are emulated by various textile methods. The optimized microscopic fiber arrangement in biological materials, finds its extension in an optimal macroscopic arrangement of struts for load-carrying structures. Figure 9.2 shows the macroscopic wood arrangement of the stem of a cactus, which found its biomimetic counterpart in the Rotex robotic arm, made manually by DLR Germany, with 0°- and 90°-fiber bundles (tensile/compression/bending forces) and 45 °-fiber bundles (torsional loads). Figure 9.3 shows the thorax of a dragonfly, in which the ribs themselves are weight-optimized structures and are macroscopically arranged as weight-saving spacer structures connected by a thin layer of outer skin.

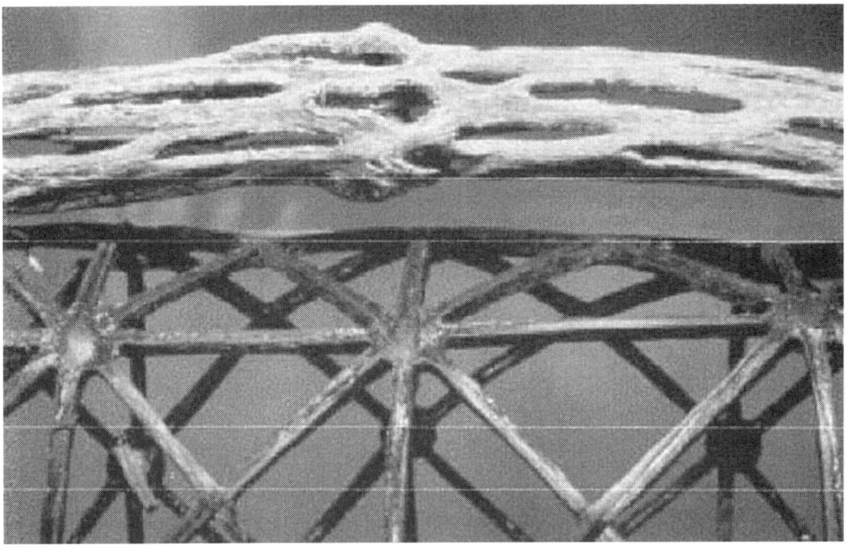

9.2 Robotic arm modeled after cactus wood (courtesy of DLR Stuttgart, Germany).

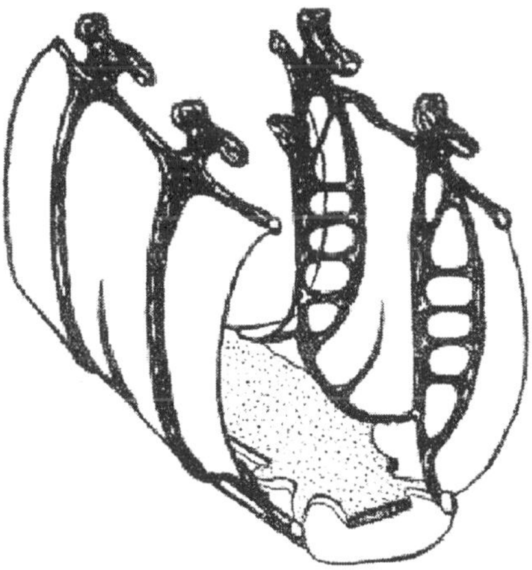

9.3 Illustration of an exoskeleton (thorax) of a dragonfly (courtesy of Prof. Nachtigall, Saarbruecken University, Germany).

A successful transfer from biological composites into technical applications dates back to the early 1980s and was accomplished by the group of R. Gordon, C. R. Chaplin, and G. Jeronimidis at Reading University. They patented a bio-inspired composite structural panel with high strength and toughness (Chaplin *et al.*, 1983) based on (ultra-)structural features in wood (Gordon and Jeronimidis, 1980; Jeronimidis, 1980, 1991). The orientation of the fibers in this biomimetic composite is based on angles found in micro-fibrils of wood tracheids.

9.2 Composites under development: 'smart composites'

Today, composites are used widely and are undergoing development. Research goes into reducing production costs, developing more sophisticated methods for constructing ultimate lightweight structures, or creating higher functionality, the functionality preferably deeply integrated into the textile or the fiber. Examples for such functionalities are: (1) passive, form-optimizing adaptive wings for maximum energy yield of airplane wings or wind turbine blades (Breitbach and Sinapius, 2004), (2) integrated (fibrous) glass sensors for damage control in bridges and airplanes, and (3) active damping of disturbing or harmful vibration with piezo-ceramic fibers (Monner, 2005).

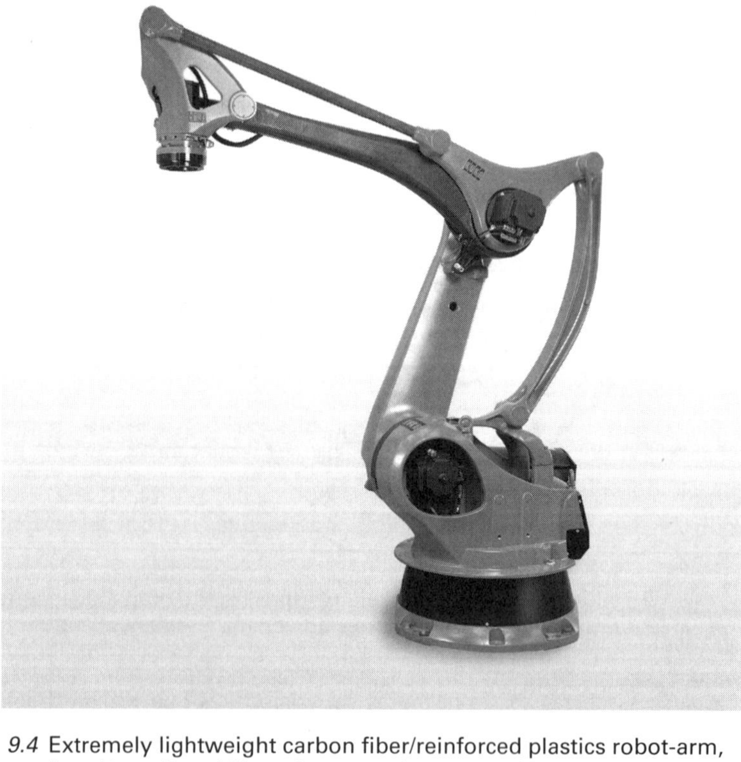

9.4 Extremely lightweight carbon fiber/reinforced plastics robot-arm, produced by tailored fiber placement (courtesy of Kuka GmbH, Germany).

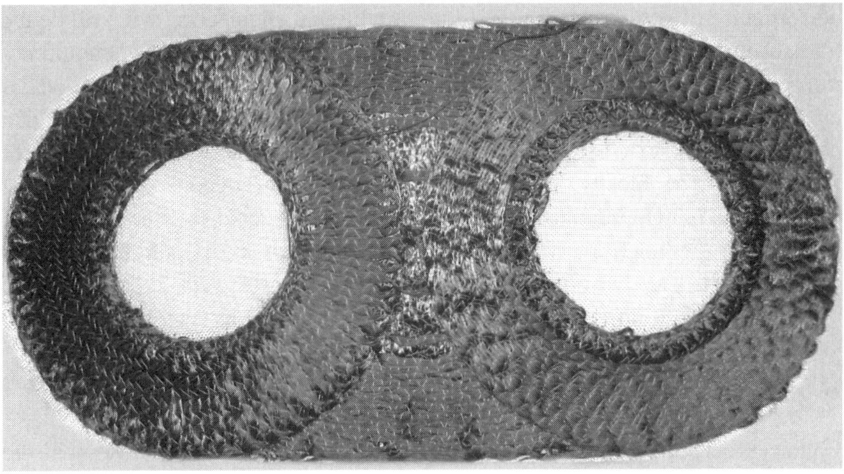

9.5 Example of a tailored fiber placement preform.

Ultimate lightweight composite structures are manufactured with so-called 'gradient textile techniques'. As in nature, every single fiber strand is exactly laid within the structure in the direction necessary to neutralize outer forces so that no unnecessary fibers or weight are incorporated. As an example for a gradient textile process, Fig. 9.4 shows an extreme lightweight carbon fiber reinforced plastic robot-arm of the German company Kuka Roboter GmbH, produced by a process called tailored fiber placement. This is a stitching process, where, on a slightly altered textile stitching machine, up to 10 stitching heads place every single carbon fiber strand next to the preceding fiber strand (Fig. 9.5).

Manufacturing of these ultra-light composites was made possible by the development of adequate 'finite element' computing methods, which facilitate the calculation of curved and irregular shapes. Together with physicists, biologists and engineers from the Forschungszentrum Karlsruhe, Claus Mattheck investigated biological design rules, triggering the development of an optimized shape of naturally growing biological constructions. They developed computing methods to simulate force-controlled biological growth, which help to optimize technical constructions by a similar 'organic' growth (as in trees) or even help to remove unnecessary volume and weight (as in bones) (Mattheck, 1990, 1996, 1998). His propagation of using tension force loaded machine elements instead of using pressure loaded elements under the motto 'Thinking in Ropes' for lightweight construction will further advance the use of custom-made composites (Mattheck *et al.*, 2004).

9.3 Using biomimetics to boost the performance of composites

For the development of fiber-reinforced plastics, biomimetics can enhance the performance of composites. In the future, traditional engineering will still be the basis of most new technical developments. Biomimetics can not and should not replace this established and well-tested approach. But new developments, whenever possible and ingenious, should be stimulated by solutions from nature and compared with nature's wisdom, thus generating a pool of ideas and knowledge for further use. The mostly superficial, functional knowledge gained from past research can now be supplemented with new findings about the fine structure of materials or the functions of boundary layers using new measuring methods.

Usually, there are two approaches to biomimetics (Speck *et al.*, 2006a, b, 2007): top-down and bottom-up, as described in the following subsections.

9.3.1 Top-down approach

Engineers are searching for ways to optimize existing products or processes with the aid of biologists and their pool of biological knowledge. Because of the differing technical languages of the researchers involved, it is very helpful if biologists acquire knowledge of various physical and engineering contexts, whereas engineers should be equally open-minded and willing to think in unusual directions. After identifying the most promising solutions with additional, and sometimes very extensive, structural and functional analyses, biologists and engineers will abstract nature, deduce the functional principles, and create a modified and appropriate technical solution. Then engineers have to look for an optimal translation into techniques with appropriate production methods and materials. This strategy is also usable for optimizing already existing biomimetic technical solutions.

9.3.2 Bottom-up approach

Biologists do fundamental research into nature's structures, processes, and functional modes of operations. New principles are discovered and analyzed, then communicated to engineers, and the two groups work together to abstract and transfer insights into new technical solutions.

Conscious that the knowledge of biological structures, processes, and functionality in plants and animals will deeply enrich engineering work, biologists from the universities of Freiburg and Tuebingen and engineers from the Institute of Textile Technology and Process Engineering in Denkendorf (ITV) founded the Competence Network Biomimetics (Speck and Speck, 2003, 2006). The interdisciplinary approach of the network members ensures that results from basic research are transferred into industrial products throughout the whole value chain. The aim of the cooperation between the biologists of the Plant Biomechanics Group and the engineers of ITV Denkendorf is the conversion of natural principles into technical applications by means of textile technologies. Textiles techniques are most suited for this purpose, because textiles techniques, like nature, also assemble from small to big, from (nanoscale-) fibres to big superstructures (Fig. 9.6).

Biological production method:

Construction from small to large:

molecule → cell → tissue → organ → living being

➤Increase in complexity by hierarchical growing

➤ **Constructing based on structures**

Traditional technical production method:

Construction from large to small:

workpiece-blank made of one material → sawing, milling etc.

➤No hierarchical structure

➤ **Constructing based on material**

Production methods based on technical textiles:

Construction from small to large:

fiber → fiber bundle → braiding, knitting etc. → textile preform → composite preform → complex formed final product

➤Increase in complexity by hierarchical composition

➤ **Constructing based on structures: similar to biology**

9.6 Comparison of biological, technical, and textile construction methods.

Because of the superior mechanical and lightweight properties of their stems, Dutch rush (*Equisetum hyemale*, Equisetaceae) (Fig. 9.7) and giant reed (*Arundo donax*, Poaceae) (Fig. 9.8) were identified by biologists from the Plant Biomechanics Group Freiburg and engineers from ITV Denkendorf as particularly promising biomimetic role models for the construction of ultra-lightweight technical structures with an interesting combination of mechanical properties (Speck and Spatz, 2001; Speck *et al.*, 2006a). With those natural examples in mind, in a joint brainstorming, biologists from Freiburg and engineers from ITV Denkendorf recombined and abstracted those natural functionalities and searched for possibilities for transferring them into technical structures.

9.4 Learning from a role model: horsetail (*Equisetum hyemale*)

The hollow aerial stem of horsetail (*Equisetum hyemale*) represents an extremely lightweight construction. Functional analyses identified a double ring structure in strengthening tissues, consisting of an outer ring of fibrous collenchymatous tissue that is connected to the inner, double-layered endodermis by 'pillar-like structures' having the appearance of T-struts in cross-section in the stem periphery

9.7 Cross-section of the stem of horsetail *Equisetum hyemale*.

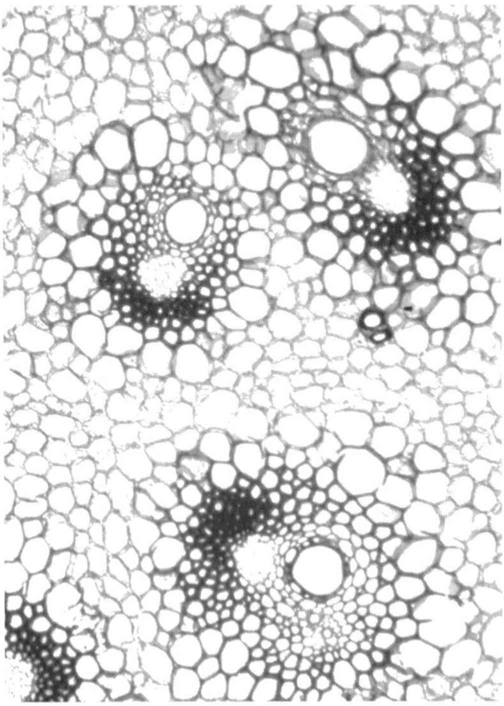

9.8 Cross-section of the stem wall of giant reed (*Arundo donax*).

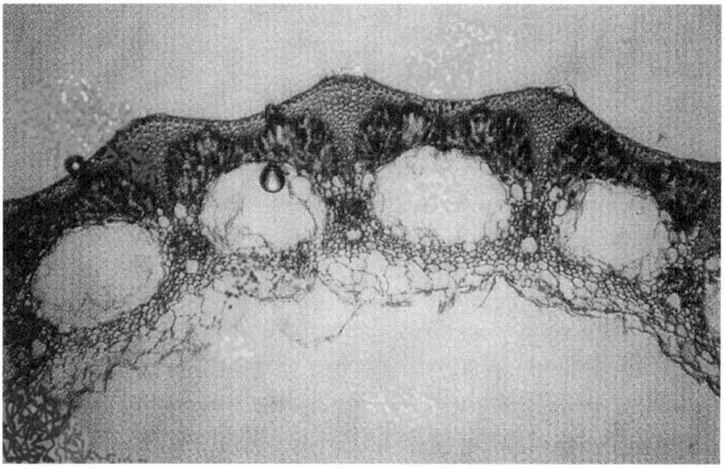

9.9 Detail of cross-section of the stem of horsetail *Equisetum hyemale.*

(Fig. 9.9). Between the collenchyma and the endodermis, which resemble a technical sandwich structure, is a thicker layer of parenchymatous tissue with remarkably large so-called vallecular canals, significantly reducing the weight of the hollow stem (Speck *et al.*, 1998; Spatz and Emanns 2004). The hollow stems of the giant reed A*rundo donax* grow up to a height of 6 m with a basal outer diameter of approximately 2 cm. They also have excellent mechanical properties under both static and dynamic loading conditions (Spatz *et al.*, 1997; Speck, 2003; Speck and Spatz, 2003). If the dense stands are subjected to dynamic wind loads, the slender culms respond with bending vibrations and pronounced damping (Speck, 2003; Speck and Spatz, 2004).

Several structural design principles, which are fundamental to theoretical mechanical engineering, contribute to these outstanding mechanical properties. The most obvious optimized structural design is the double ring structure of *Equisetum hyemale*. This principle is well known in mechanical engineering. In load-carrying beams, most of the material of the beams should be placed (or spaced) at the utmost possible distance from the center of the beam (neutral line), this being the reason for developing the double T-beams. The structure of Equisetum hyemale anticipated one of the most applied structural principles in building bridges and houses and just looks like welded together double T-beams (Fig. 9.10).

The same principle is also applied in building cars (space frame structure) or airplanes. In airplanes, the honeycomb or foam cores in 'spacer' sandwich composites multiply the bending resistance of structures. Figure 9.11 illustrates this principle. The left, thin side of the specimen is comprised of two layers of glass woven fabric embedded in a matrix material and has a bending resistance defined as 1. In the thicker, right side of the specimen, a foam core is included between the

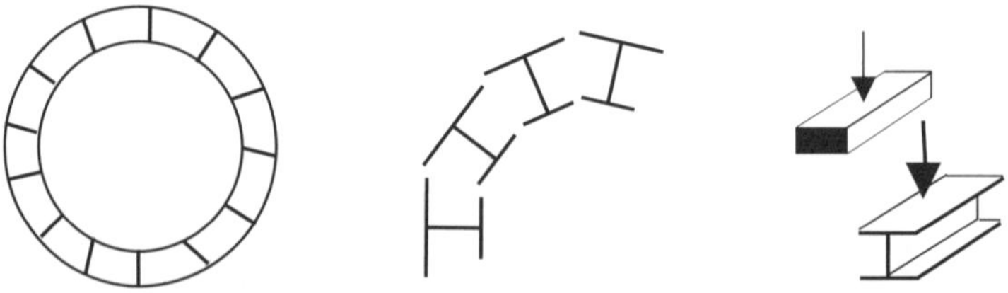

9.10 Double-ring structure with connecting beams, similar to *Equisetum hyemale,* 'constructed' out of welded double T-beams.

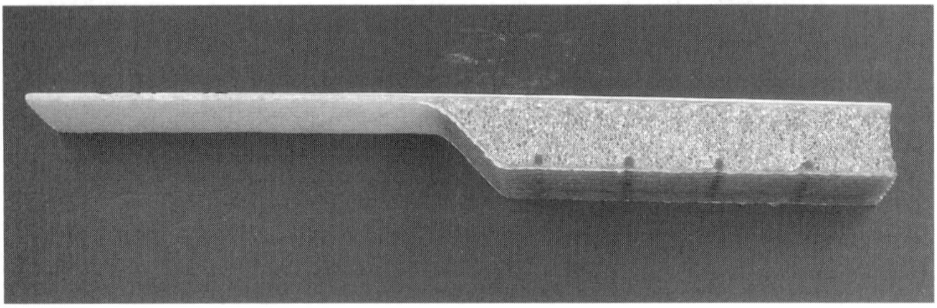

9.11 Increasing bending stiffness of composites using a foam core.

9.12 Cut-away view of an 'Isogrid' bicycle frame tube to show inner stitched fiber core reinforcement (courtesy Vyatek Corp.).

skin layers. This part has a bending moment 40 times higher than the left side without the core.

For the same reason, hollow tubes also have very good specific bending resistance. But, if the skin of the tube is very thin, buckling will arise, and the structure will need a supportive inner framework, either a foam core or a fibrous

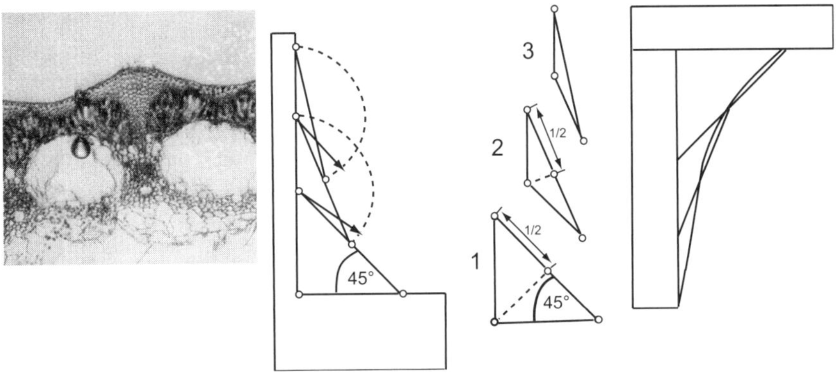

9.13 Optimally shaped connection between pillars and rings, shaped similarly as with trees to avoid notch stress (courtesy of Prof. Mattheck).

core (Niklas, 1989, 1992, 1997; Mattheck, 1996, 1998, 2006; Spatz and Speck, 1994; Spatz *et al.*, 1997; Speck *et al.*, 1998; Mattheck *et al.*, 2006).

9.4.1 First bionic transfer

With its double-ring structure with connecting cross-beams, *Equisetum hyemale* represents a superior lightweight construction with high compression and bending stiffness. So far, fibrous cores in tubular composites are very seldom applied, and never before in pultruded tubular composites. However, Fig. 9.12 shows a newly developed lightweight tubular composite material, where fiber strands are stitched helically onto the inner side of a thin tubular skin, the part being yet very expensive. The nearly exact structure of *Equisetum hyemale* can be produced with braid-pultrusion technology, the basic equipment already installed at ITV Denkendorf. For producing a similar effective structure the braid-pultrusion line was specially adapted.

9.4.2 Second bionic transfer

The curved shape of the cross-beams of *E. hyemale* offers a continuous and organic connection with better stress distribution between the inner and outer ring than analogous technical constructions with a right angle bonding. Thus notch stress is largely avoided. One can see this principle just everywhere in nature, most notably in trees with specially adapted, shape-optimized connections from trunk to branches and to the root (Mattheck 1990, 1996, 1998). In the technical realization, the connection between pillars and rings will be shaped similarly (Fig. 9.13).

9.4.3 Third bionic transfer

Vallecular canals between inner and outer ring and the connecting cross-beams of

E. hyemale work as an additional means of gas exchange and store water from the cells to prevent frost damage. As technical analogs of the vallecular canals, the so-called functional canals could be used to transport electrical power via power supply lines or liquids and gases via pipes (see Fig. 9.9).

9.5 Learning from a role model: giant reed (*Arundo donax*)

The hollow stems of *Arundo donax* offer another astonishing sort of 'core': a sophisticated, weight-optimized structure with a material optimized to dynamic loads. The stems are composed of strengthening elements such as vascular bundles with accompanying fiber caps, which are embedded in a matrix of basic parenchyma (Fig. 9.8). In cross-section, at least four structural gradients on different hierarchical levels can be found, which meet all theoretical considerations and needs of composite structures.

In the periphery of the hollow stems, the area of highest stress, most of the load-carrying fiber material is placed. Then, the amount of load-carrying material is gradually reduced in the direction of the stem's hollow pith, in keeping with the gradual decrease in bending stress as the distance from the periphery increases.

This mechanical grading is also exemplified by the lignification of the parenchymatous basic tissue as it decreases in a radial direction from the outside toward the center.

An additional gradient in the basic parenchyma can be found: the increasing size of the parenchyma cells is accompanied by gradually thinner cell walls from the outside to the inside of the stem wall, causing a reduction of the relative cell wall amount.

The macro superstructure finds its counterpart in a micro-structural phenomenon. The pronounced difference in stiffness between natural fibers and the surrounding parenchyma matrix is equalized by a gradual transition in stiffness. This results in a very high damping of oscillating wind forces and a high bending ability with optimal distribution of stress before the connection between fibers and matrix finally fails and the structure disintegrates. This is illustrated by Fig. 9.14, which shows a bending test of an internode of the hollow stem of *Arundo donax*. After each of the 10 smaller ruptures, the structure stabilizes itself through the distribution of stress and can then tolerate even more stress until the final failure. This graph is a very good example of a mechanically benign, ductile rupture failure and could be a model for any load-carrying technical structure (Spatz *et al.*, 1997; Speck *et al.*, 2006a).

9.5.1 Fourth bionic transfer

According to the gradually decreasing bending stress with increasing distance from the outer periphery of the plant stem, the amount of load-carrying material is

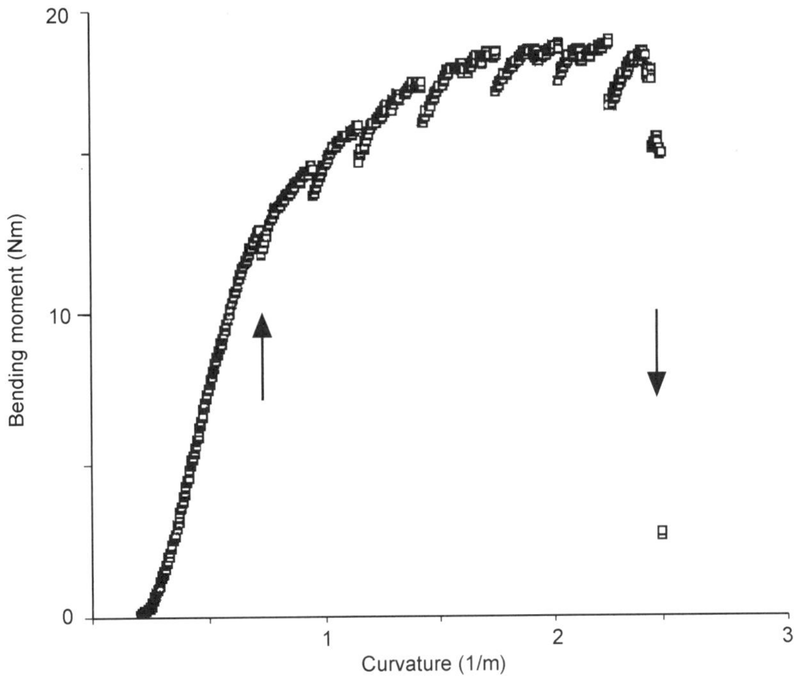

9.14 Bending test of an internode of *Arundo donax* to show the relationship between bending moment and curvature. Arrows mark initial failure and final failure. In between, a series of several partial collapses can be found.[19]

also gradually reduced concurrent with a gradual reduction of cell wall thickness and an increase of cell size of the parenchymatous basic tissue.

As 'gradient textile technologies' already demonstrate, fibers should only be incorporated where they are useful in carrying loads. In less stressed parts of a construction, polymer foam will be used for spacing out fibers. With the braid-pultrusion technique, this principle can be applied very easily into the structure.

9.5.2 Fifth bionic transfer

The angle of strengthening fibers in the stem walls and of cellulose micro-fibrils in cell walls is optimized according to the types and combinations of mechanical loading occurring in the plant stem (e.g. static or dynamic bending and torsion, energy damping). The angle of fibers and fiber bundles can be adjusted to the predominant loading situation(s) of the application mode of the finished component by using computer-controlled braiding techniques available at ITV Denkendorf. Optimally designed technical structures can thus be produced.

9.5.3 Sixth bionic transfer

In summary, the structural basis of graded transition of Young's modulus between stiff fiber and less stiff parenchyma matrix is a gradient in lignification between fibers and parenchymatous cells and of variations in cell size and cell wall thickness. Transferring this knowledge into techniques, the stiffness between fibers and matrix will also be graded using a matrix material that can be provided by varying Young's modulus.

9.6 Learning from a role model: wood

Another interesting structure can be found in wood (Fig. 9.15). Helically arranged cellulose micro-fibril bundles, as found for example in the different layers of conifer tracheids, may result in a diagonally 'braided' structure improving (torsional) stability and oscillation damping (Mark and Gillis, 1973; Cave and Walker, 1994; Reiterer *et al.*, 1999; Burgert *et al.*, 2004, 2005).

9.6.1 Seventh bionic transfer

Tracheids with spirally arranged micro-fibril bundles improve torsional stability and oscillation damping. Using the braiding technology developed at the ITV Denkendorf, helically wound fibers can be easily incorporated into the structure.

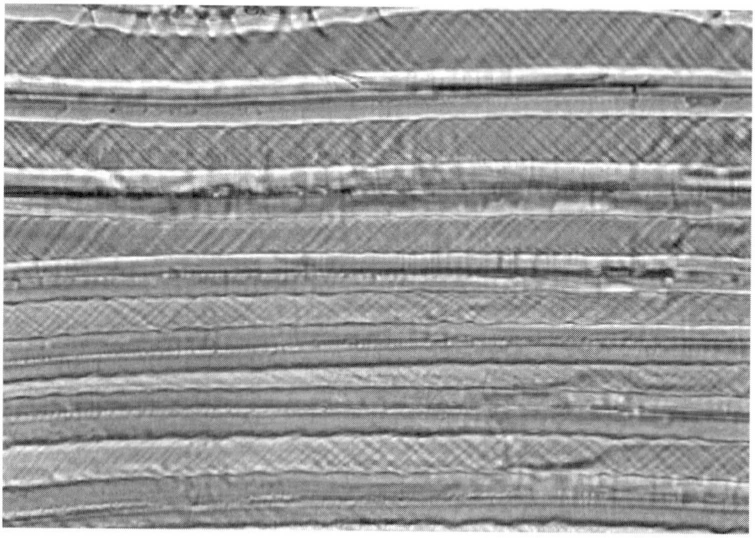

9.15 Polarized light microscopy of compression wood of spruce (*Picea abies*), image of helical cellulose fibers in adjacent cell walls (courtesy of Dr I. Burgert, Max Planck Institute of Colloids and Interfaces, Potsdam/Golm, Germany).

9.7 Combination of different principles of role models into the 'technical plant stem'

9.7.1 Eighth bionic transfer: creating the 'technical plant stem'

Whereas other bionic solutions in the field of composites are mostly founded on a single natural function, the above-mentioned seven different natural functionalities were combined and translated into one technical structure, creating a totally new technical fiber composite material with superior mechanical properties.

The advantages of the new composite material were self evident in such a way that specialists from prominent composite companies – seeing first prototypes of the technical plant stem – encouraged the inventors to patent it. The now-patented material, called the 'technical plant stem', has generated interest from several composite companies from the fields of aerospace, automotive, building, and sporting goods, which are willing to finance future research work.

9.8 Production methods and machinery equipment for the 'technical plant stem'

Nevertheless, the production costs for combining these functionalities into a new, unique product, must always be considered. For composite beams, the pultrusion process is a cost-efficient production method for endless-fiber-reinforced plastics. Compared with metals, the profiles are corrosion resistant and to a large degree maintenance free. They are very safe in having good electrical and thermal isolation, installation costs are lower and lighter foundations can be realized. In thermo-set pultrusion, impregnated high-performance fibers are pulled through a form-shaping die and are consolidated by heat and pressure during the transit through the die (Fig. 9.16). To incorporate diagonal fiber bundles into the technical plant stem, a braiding machine was installed into the pultrusion process (Fig. 9.17 and 18).

The braiding technique (Fig. 9.19) helically winds two different counter-rotating sets of intertwining fiber strands around a core system and an inner layer of unidirectional fibers. Deriving from ancient arts, braiding is now a process with ever-increasing fields of applications, such as composites, technical, or smart textiles. By varying the density, arrangement, and angles of the fibers in the different layers of the technical plant stem, technical structures optimally designed for a given load situation can be produced. The targeted structure and line production of the technical plant stem are only feasible with computer-controlled braiding line equipment.

In a first approach, different hollow profiles with glass fiber reinforcement were pultruded and braid-pultruded, optimizing the process and machinery (Fig. 9.20 and 21).

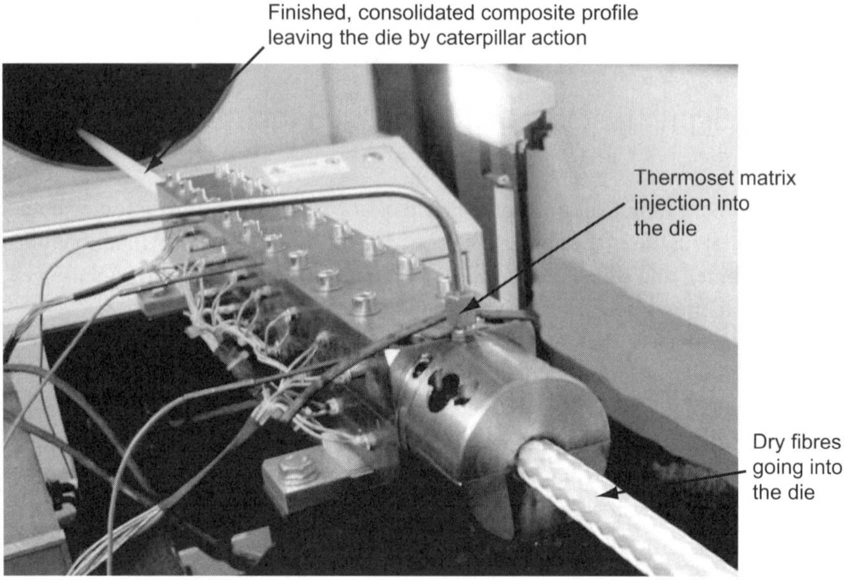

9.16 ITV thermoset braid-pultrusion.

9.17 ITV braid-pultrusion: (1) reel for core- and axial-yarns, (2) braider with braiding yarns, (3) preheating device, (4) heated die, (5) water cooling, (6) caterpillar.

Parallel to this process, first samples of technical plant stems were braided and resin impregnated (Fig. 9.22). In the next step of optimization, the usual polyurethane matrix was used to encase micro-fibrils and fiber bundles. But to mimic the porous, optimized weight potential of the plant–matrix system, a polyurethane foam matrix was applied between the fiber bundles, resulting in a very lightweight

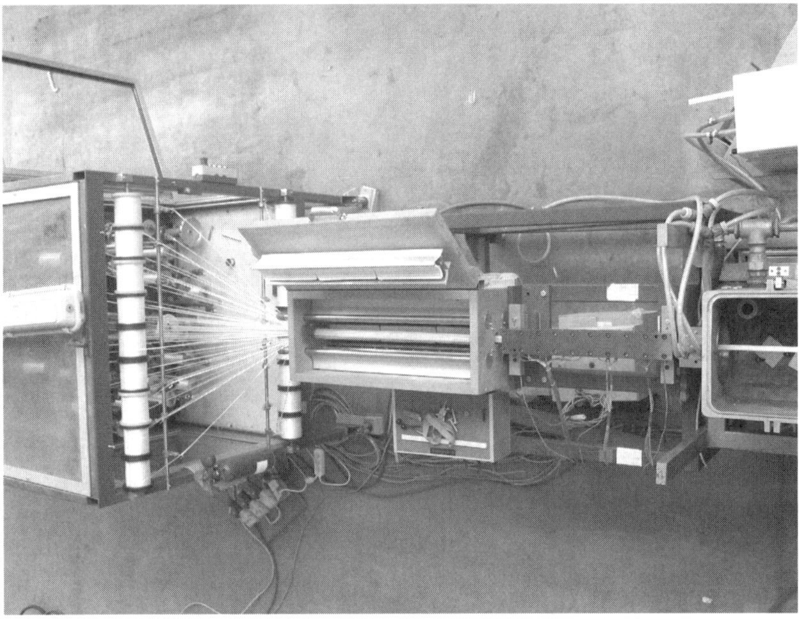

9.18 ITV braid-pultrusion system, seen from above.

9.19 Detail of braiding technique.

9.20 Thin-walled, unidirectional, reinforced thermoplastic profile produced by pultrusion.

9.21 Braid-pultruded tubular profile with thermoset matrix.

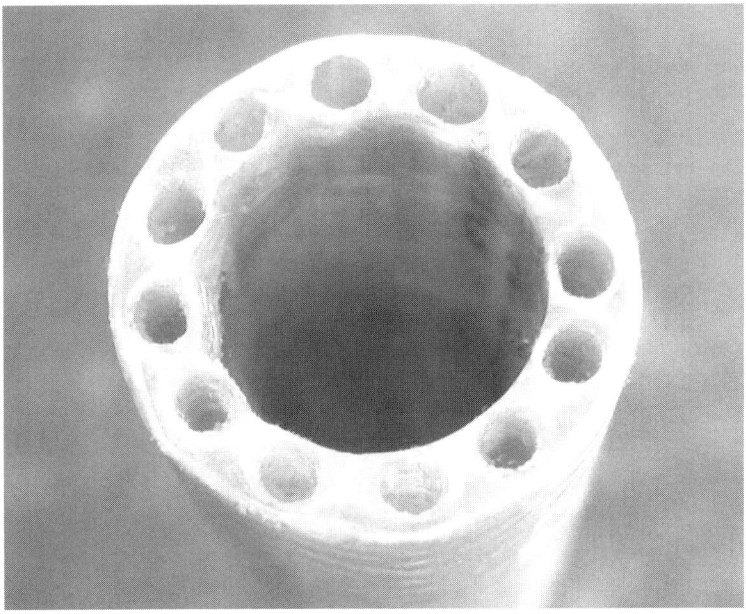

9.22 'Technical plant stem' with double-braid textile construction.

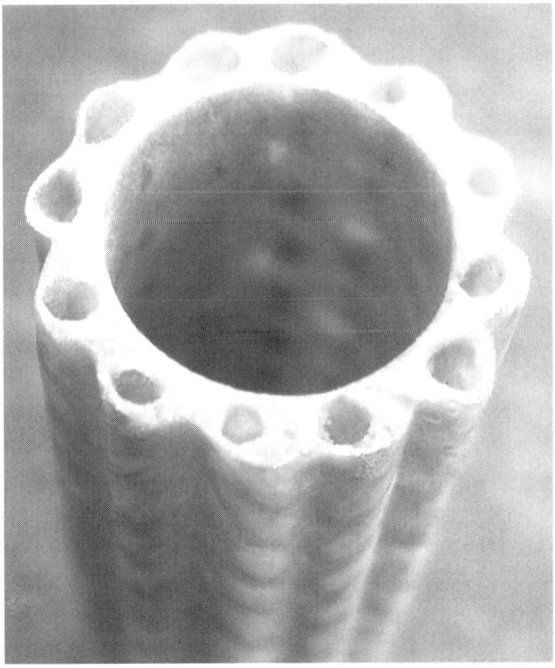

9.23 'Technical plant stem' with partial polyurethane foam matrix.

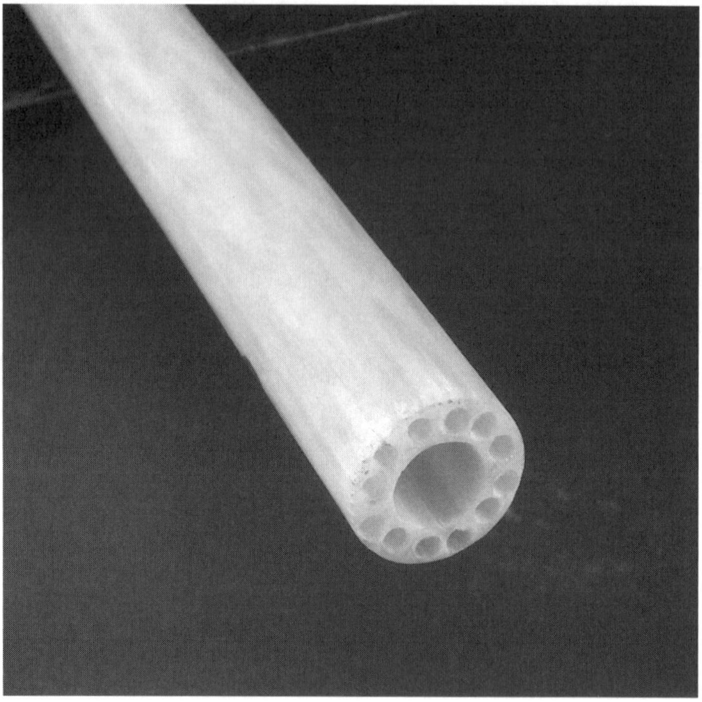

9.24 Braid-pultruded 'technical plant stem.'

specimen (Fig. 9.23). The first sample of a braid-pultruded technical plant stem can be seen in Fig. 9.24, showing the principle applicability of the braid-pultrusion process. By combining this cost-efficient production method with the several advantages of the technical plant stem, this unique material will rapidly spread into manifold applications in the composite world.

9.9 Applications of the 'technical plant stem'

There are numerous applications for the technical plant stem. Usually, where tubular rods are used – and there is an abundance of applications for tubular rods – the structure of the technical plant stem will enhance the performance of the original rod regarding mechanical properties like pressure resistance or bending stiffness. Either the rods will be stiffer and stronger, or the rods will be lighter than other composite materials.

In the building industry, if the rods are used to build a load-carrying structure under the roof, the structure is assembled very rapidly with less manpower and the foundations are made lighter (Fig. 9.25). This is the same as in telescopic structures, which need a stiff yet lightweight construction, because the whole structure will be moved in a desired direction.

9.25 Application of the 'technical plant stem': lightweight load-carrying roof substructure (courtesy Exel-Oy).

In other instances, the multi-channel system will offer new functionalities like an integrated gas, fluid or power transport, or using the structure as a pre-tensioned system.

9.10 Conclusions and future trends

Many other astonishing plant functionalities are waiting to be discovered. The transdisciplinary collaboration of researchers with biological, chemical, physical and engineering backgrounds and a systematic biomimetic approach to comprehend biological structures, processes, and functionality will bring new insights for the development of new technical solutions. In our opinion, technical textiles and fibrous composite materials offer a brilliant opportunity for transferring ideas inspired by biological models via biomimetic approaches into innovative technical structures, because composite materials based on technical textiles allow production processes similar to those used by nature (Milwich *et al.*, 2006, 2007).

For composites, the way into the future is already predetermined by nature: combining the lightweight nature and energy-saving potential in production and use of composites with a better recycling ability. Improving the recycling ability e.g. with the use of biodegradable natural fibers and bio matrix systems, the future use of composites will quickly increase. The development of the technical plant stem will strongly contribute to future developments.

9.11 Acknowledgement

The authors gratefully acknowledge the support of the research project by the Competence Network 'Plants as Concept Generators for Biomimetic Materials and Technologies' of the State of Baden-Wuerttemberg, Germany.

9.12 References

Breitbach E. and Sinapius M. 2004. Stand und Perspektiven des Leichtbaus und der Adaptronik. Jahresbericht des DLR Instituts fuer Strukturmechanik, Braunschweig, Germany.

Burgert I., Fruehmann K., Keckes J., Fratzl P. and Stanzl-Tschegg S. E. 2004. Structure-function-relationships of four compression wood types – Micromechanical properties at the tissue and fiber level. *Trees – Structure and Function* **18**: 480–485.

Burgert I., Gierlinger N. and Zimmermann T. 2005. Properties of chemically and mechanically isolated fibres of spruce (*Picea abies* [L.] Karst.). Part 1. Structural and chemical characterization. *Holzforschung* **59**: 240–246.

Cave I. D. and Walker J. C. F. 1994. Stiffness of wood in fast-grown plantation softwoods: the influence of micro fibril angle. *Forest Products Journal* **44**: 43–48.

Chaplin C. R., Gordon J. E. and Jeronimidis G. 1983. United States Patent. Composite Material. Assignee: Westvaco Corporation, New York, USA, Application 351,777.

Gordon J. E. and Jeronimidis G. 1980. Composites with high work of fracture. *Philosophical Transactions of the Royal Society of London*, **A 294**: 545–550.

Harder D. (ed.) 2006. BIOKON Bionik-Kompetenz-Netz – Creative transfer of biological principles into engineering. BIOKON e.V. Bionics Competence Network, Berlin, Germany.

Jeronimidis G. 1980. Wood, one of nature's challenging composites. In J. F. V. Vincent and J. D. Currey (eds.), *The mechanical properties of biological materials, Symposia of the Society for Experimental Biology*, **34**: 169–182. Cambridge University Press, Cambridge, UK.

Jeronimidis G. 1991. Learning from nature: biological composites. Proceedings of Inaugural European Seminar, Arche de la Defense, Paris, France.

Lauterbach C. 2005. ADNOS – ein selbstorganisierendes Sensornetzwerk fuer Smart Textiles. Proceedings of Kolloquium 'Smart Textiles – Vom Rohstoff bis zum Endprodukt', ITV Denkendorf, Germany, lecture no. 18.

Linti C. and Horter H.-J. 2005. Baby-Body mit Sensorik zur Erfassung von Vitalparametern. Proceedings of Kolloquium 'Smart Textiles – Vom Rohstoff bis zum Endprodukt', ITV Denkendorf, Germany, lecture no. 16.

Mark R. E. and Gillis P. P. 1973. The relationship between fiber modulus and S2 angle. *Tappi* **56**: 164–167.

Mattheck C. 1990. Engineering components grow like trees. *Materialwissenschaft und Werkstoffkunde* **21**: 143–168.

Mattheck C. 1996. *Trees – the mechanical design*. Springer Verlag, Heidelberg, Germany.

Mattheck C.1998. *Design in nature – learning from trees*. Springer Verlag, Heidelberg, Germany.

Mattheck C., Kappel R., Tesari I. and Kraft O. 2004. In Seilen denken – Einfache Anleitung fuer Naturnahes Konstruieren. *Konstruktionspraxis* **9**: 26–29.

Mattheck C., Bethge K. and Tesari I. 2006. Shear effects on failure of hollow trees. *Trees – Structure and Function*, **20**: 329–333.

Milwich M. and Mueller E. 2005. Medical textiles for implants. Proceedings of 'Germany meets India', Seminar on Technical Textiles, Mumbay, Federation of All India Textile Manufacturers' Associations (FAITMA), India.

Milwich M., Speck T., Speck O., Stegmaier T. and Planck H. 2006. Biomimetics and technical textiles: solving engineering problems with the help of nature's wisdom. *American Journal of Botany*, **93**: 1295–1305.

Milwich M., Planck H., Speck T. and Speck O. 2007. The technical plant stem: a biomimetically inspired narrow fabric. *Melliand – Narrow Fabric and Braiding Industry*, **44**: 34–38.

Monner H. P. 2005. Smart materials for active noise and vibration reduction. *Proceedings of the Novem – Noise and Vibration: Emerging Methods*, Saint-Raphael, France.

Niklas K. J. 1989. Nodal septa and the rigidity of aerial shoots of *Equisetum hyemale*. *American Journal of Botany* **76**: 521–531.

Niklas K. J. 1992. *Plant biomechanics. An engineering approach to plant form and function*. University of Chicago Press, Chicago, Illinois, USA.

Niklas K. J. 1997. Responses of hollow, septate stems to mechanical vibrations: evidence that nodes can act as spring-like joints. *Annals of Botany* **80**: 437–448.

Reiterer A., Lichtenegger H., Tschegg S. and Fratzl P. 1999. Experimental evidence for a mechanical function of the cellulose microfibril angle in wood cell walls. *Philosophical Magazine* **A 79**: 2173–2184.

Spatz H.-Ch., Beismann H., Bruechert F., Emanns A. and Speck T. 1997. Biomechanics of the giant reed *Arundo donax. Philosophical Transactions of the Royal Society of London, B, Biological Sciences* **352**: 1–10.

Spatz H.-Ch. and Emanns A. 2004. The mechanical role of the endodermis in *Equisetum* plant stems. *American Journal of Botany* **91**: 1936–1938.

Spatz H.-Ch. and Speck T. 1994. Local buckling and other modes of failure in hollow plant stems. *Biomimetics* **2**: 149–173.

Speck O. 2003. Field measurements of wind speed and reconfiguration in *Arundo donax* (Poaceae) with estimates of drag forces. *American Journal of Botany* **90**: 1253–1256.

Speck O. and Spatz H.-Ch. 2003. Mechanical properties of the rhizome of *Arundo donax* L. *Plant Biology* **5**: 661–669.

Speck O. and Spatz H.-Ch. 2004. Damped oscillations of the giant reed *Arundo donax* (Poaceae). *American Journal of Botany* **91**: 789–796.

Speck O., Milwich M., Harder D.L. and Speck T. 2005. Vom biologischen Vorbild zum marktreifen bionischen Produkt: der 'technische Pflanzenhalm'. *Museo* **22**: 96–103.

Speck T. and Harder D. 2006. Bionics or biomimetics: taking a leaf out of nature's book. In D. Harder (ed.), BIOKON Bionik-Kompetenz-Netz–Creative transfer of biological principles into engineering, 4–6. BIOKON e.V. Bionics Competence Network, Berlin, Germany.

Speck T., Harder D., Milwich M., Speck O. and Stegmaier T. 2006a. Die Natur als Innovationsquelle. In P. Knecht (ed.), *Technische Textilien*, 83–101. Deutscher Fachverlag, Frankfurt, Germany.

Speck T., Harder D. and Speck O. 2006b. BIOKON centers in brief – Freiburg. In D. Harder (ed.), BIOKON Bionik-Kompetenz-Netz–Creative transfer of biological principles into engineering, 42–43. BIOKON e.V. Bionics Competence Network, Berlin, Germany.

Speck T., Harder D. and Speck O. 2007. Gradient materials and self-repair: learning technology from biology. In *Plastics in Automotive Engineering*, 1–13. VDI Wissensforum, IWB GmbH, VDI-Gesellschaft Kunststofftechnik, VDI Verlag GmbH, Düsseldorf.

Speck T. and Neinhuis C. 2004. Bionik, Biomimetik. *Naturwissenschaftliche Rundschau* **57**(4): 177–191.

Speck T. and Spatz H.-Ch. 2001. Transkription oder Translation: Pflanzen als Ideengeber fuer neue Materialien und technische Leichtbaustrukturen. In A. von Gleich (ed.), *Bionik*, 2nd ed., 229–245. Teubner Verlag, Stuttgart, Germany.

Speck, T. and Speck O. 2003. Competence network biomimetics: Taking a leaf out of nature's book. *The Official Journal of the BioValley Network* **1**: 23–24.

Speck, T. and Speck O. 2006. Eng verflochten und gut gestrickt: Das Kompetenznetz Biomimetik. Proceedings of the 1. Bionik-Kolloquium: Bio-inspired Textile Materials: 18 pp, Institut für Textil- und Verfahrenstechnik (ITV) Denkendorf (published on CD).

Speck T., Speck O., Emanns A. and Spatz H.-Ch. 1998. Biomechanics and functional anatomy of hollow stemmed sphenopsids: II. Equisetum hyemale. *Botanica Acta* **111**: 366–376.

Stegmaier T., Milwich M., Scherrieble A., Geuer M. and Planck H. 2004. Bionik developments based on textile materials for technical applications. In I. Boblan and R. Bannasch (eds.), Fortschritt-Berichte VDI, Reihe 15 Umwelttechnik, 249: 323–330. VDI Verlag GmbH, Duesseldorf, Germany.

Stollbrock O. 2005. mp3blue – die Multimedia Lifestylejacke. Proceedings of the Kolloquium 'Smart Textiles – Vom Rohstoff bis zum Endprodukt', ITV Denkendorf, Germany, lecture no. 12.

Vogel S. 1998. *Cats' paws and catapults – mechanical worlds of nature and people*. Norton, New York, USA.

10

Bionic developments based on textile materials for technical applications

T. STEGMAIER, V. VON ARNIM, M. LINKE, M. MILWICH, J. SARSOUR, A. SCHERRIEBLE, P. SCHNEIDER and H. PLANCK
Institute of Textile Technology and Process Engineering Denkendorf (ITV), Germany

Abstract: Developments in the production of self-cleaning textiles based on the 'Lotus Effect' are presented and topical bionic research activities being undertaken in this area at ITV Denkendorf are described. The biological models have been analyzed in detail to allow definition of the construction requirements and tube-filter development, assembly of filtration test equipment and determination of characteristic filtration data. A general model for lightweight tubular structures with high bending stiffness was defined, and this formed the basis for studying potential industrial applications, including solar-thermal materials that mimic the pelts of polar bears for use in energy technology.

Key words: biomimetics, bionic textiles, nano technology, self-cleaning surfaces, 'Lotus Effect'.

10.1 Introduction

Fiber technologies based on natural models of growth processes, hairy structures and reinforcement show great potential for the development of bionic materials. Biologists and engineers have conducted intensive bionic research and development activities at ITV Denkendorf for about eight years, and research areas include:

- surface technology, exploring possibilities for self-cleaning surfaces and boundary layers that retain air and reduce friction in water, based on the lotus leaf;
- environmental technology, studying self-adapting filtration systems as potential models for energy-independent liquid transport;
- energy technology, developing new solar–thermal materials that mimic the behavior of polar bear pelts;
- climate control through the use of adaptive and breathable membranes; and
- lightweight construction products with load-bearing properties based on fiber orientation and micro-macro-graduation created by pultrusion.

10.2 Potential of fiber-based materials in bionics

The term 'bionics' encompasses the areas of biology and technology, and describes the creative transfer of findings from the living world of nature into technical products and systems. An intensive and interdisciplinary cooperation between natural scientists (e.g. biologists) and engineers/technicians is essential for this type of research.

10.2.1 From micro- to macro-structures

Technology tends to start with large volumes of raw materials, which are gradually processed into smaller functional units and subsequently assembled, whereas nature does the opposite. Genetic information and environmental influences control growth processes that start with the smallest units – atoms and molecules – and generate larger structures. This growth can result in very complex systems with functionality and efficiency that far exceeds any of our technical products, especially in terms of consumption of materials and energy.

Textile-processing technologies offer remarkable analogies to natural growth processes. Starting from small units of single fibers, down to nanometer dimensions, larger elements can be 'composed' in staged processes. This method in principle functions without producing much waste and requires relatively small amounts of energy.

10.2.2 Hairy surfaces

Many surfaces in nature are hairy. Hairs can be found on the upper and under sides of insects, between the parts of an insect exoskeleton, in the feathers of birds, in the coats of animals, and in spiders' webs. These surfaces have functions that we know a lot about, but we are far from knowing everything. The many different methods of fiber processing, fiber orientation and finishing are suitable for meaningfully transferring biological functions to technical products.

10.2.3 Fibers in composites

Nature has many forms of fiber-reinforced materials. The more closely these materials are analyzed, the more astonishing the findings are. Fibers at the nanoscale, gradual transitions, high-tensile materials, and functional cross-sections can be found in natural constructions.

Composites occur in soft and hard forms in bones, stalks, leaves, and surfaces, and are composed of organic and inorganic materials.

Topical bionic R&D activities being undertaken in these areas at ITV Denkendorf are described below.

10.1 Spherical water drops on fabric with Lotus Effect.

10.3 Research activities in the field of surfaces

10.3.1 Self-cleaning surfaces according to the model of lotus leaves

Self-cleaning textiles based on the 'Lotus Effect'[1] are being developed at ITV Denkendorf using our understanding of the basic correlations between textile constructions, surface topographies and wetting behavior after water-repelling functionalization.

Previous investigations, for example, have shown that wetting behavior is considerably influenced by topography, which results from weave and type of yarn. The type of yarn – staple fiber or filament yarn – determines the extent to which water drops roll off textiles. Long, protruding fibers, which, for example, occur with increasing frequency in ring yarn samples, prevent water drops from rolling off. Short, protruding fibers at high fiber density, as for example in the case of nonwovens, promote rolling off because they present a smaller surface area to the water drops (Fig. 10.1). Self-cleaning ability is determined by the ability of water drops to access dirt particles. Thus, good water-repellent behavior, which results from the hairiness of the sample, does not automatically lead to good self-cleaning behavior. Lotus leaves exhibit a hydrophobic micro- and nano-structure

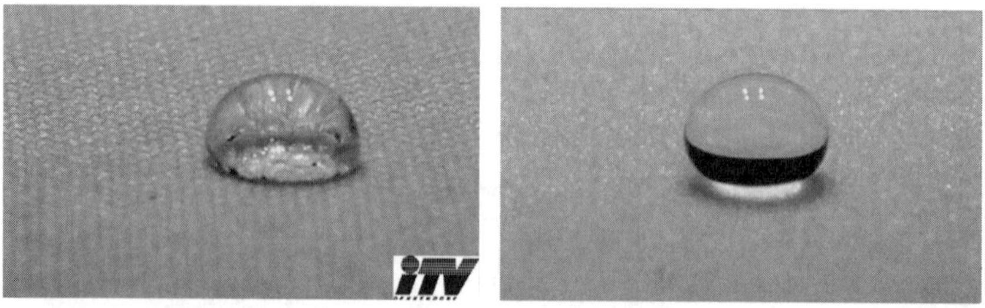

(a) (b)

10.2 Water drop on hydrophobic, flocked textiles with (a) low and (b) high flock density.

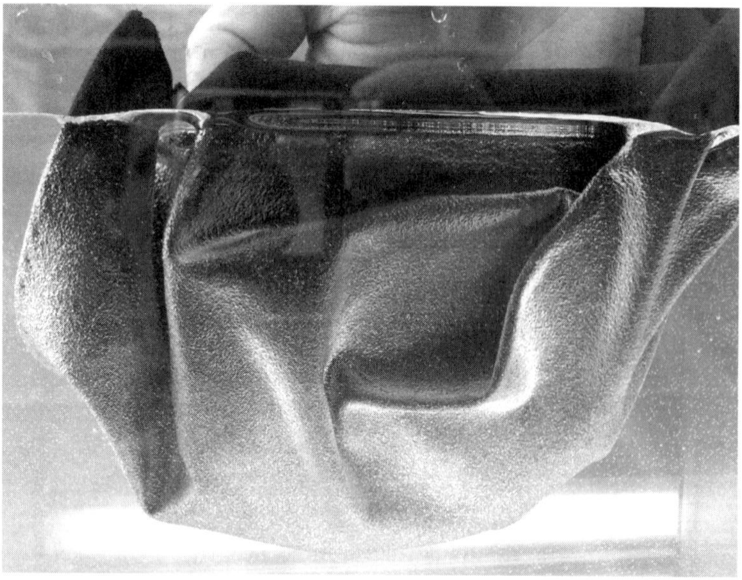

10.3 Textile made of bulky special yarn with especially good air-storing property.

without any hair which leads – due to superhydrophobicity – to outstanding self-cleaning properties. The opposite is the case with lady's mantle, where the hairs are located above the hydrophobic surface of the leaf. These are able to keep the rain drops away from the surface of the leaf, but prevent smaller drops from rolling off.

We are now at the stage where a seal of approval for self-cleaning textiles based on the Lotus Effect can be awarded.

10.3.2 Air-storing boundary layers under water

In order to prevent wetting, plants and animals use surfaces with hydrophobic, hairy coverings. Nature pursues several strategies that permit birds and animals to swim and dive without getting wet, but hairy surfaces are the most distinct. Textiles, with their fiber-based construction, have a large potential for developing boundary layers with lasting air-storing properties. Textiles with hairy surfaces are flocked. The transition from a hydrophobic but wettable surface of low flock density to a non-wettable, superhydrophobic surface of high flock density – where a stable air layer forms between drop and flocked substrate – is illustrated in Fig. 10.2.

Investigations have shown that optimized textile surfaces generated from particularly bulky, hydrophobically finished special yarns remained dry after up to four days' storage in water (Fig. 10.3).

10.4 Research activities in environmental technology

10.4.1 Self-adapting micro-filtration systems

Adaptive filtration plays an important role in nature, especially for food ingestion in sponges (Fig. 10.4). Water is sucked into the sponge through fine pores placed over the entire surface, food particles and oxygen are extracted from the water and then it is ejected via larger, collective channels. The water stream is driven by special cells called choanozytes, which are located in groups along the pore

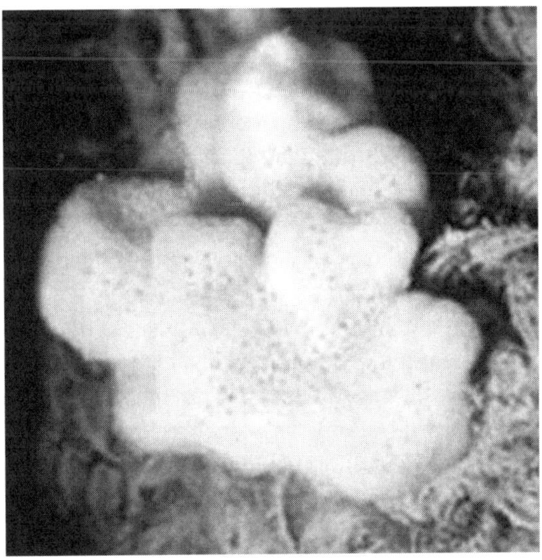

10.4 Cushion-shaped sponge.

channels. Freely mobile cells (amoebozytes) take up the food in the interior of the sponge and distribute it to the rest of cells, which are incapable of nourishing themselves independently. The sponges can change the pore size of the channels using a tissue similar to muscle, so the water stream and pressure ratios can be adjusted. This method of filtration is extremely energy-efficient, and effective pressures from 10^{-4} to 10^{-5} bar result in up to half of the body volume in water being moved per second.[2, 3]

What nature performs so successfully is difficult to replicate, but low-energy and efficient separation processes that worked reliably over a long lifetime and did not clog would be highly valuable in the field of micro-filtration, so we initiated a feasibility study. The first step was to use fiber innovations to develop an adaptable and low-energy cross-flow micro-filtration system. A specially braided tube filter with pore sizes that adapted over time was developed. Twined monofilaments were found to increase the filtration efficiency due to their micro-roughness, a considerable advantage for surface cleaning.[4]

Examinations and tests performed

The biological models were analyzed in detail, and the following steps were carried out:

- definition of the construction requirements and tube filter development;
- assembly of filtration test equipment;
- determination of characteristic filtration data.

Construction requirements and tube filter development

Filtration took place through the tube cover in one braided tube filter prototype. In principle, filtration from the outside to the inside and vice versa is possible, corresponding to requirements. The innovation here was to use twisted monofilament fibers instead of the normal multifilament fibers of the polyamide (PA) fiber material studied. The helical fiber contour with no round form results in micro-porosity along the partially parallel twisted contours in the wickerwork (Fig. 10.5). The additional micro-porosity produces an increased pore surface, which facilitates filtrate flow and reduces the pressure losses substantially.

Bending resistant and twisted monofilaments make the tube filter more stable, and the braid form is significantly more stable against external pressure. This is very favorable, especially in the case of external to internal filtration, when filtration can take place under a significantly increased pressure difference. Introducing a stranded or twisted core into the braid provides additional support for the braid tube under higher pressures.

The filter effect can be modified by adjusting the tube length, which defines the angle of the braid fibers to each other and thus the pore size. Fig. 10.6 shows the geometry for pore size calculation of a cylindrical braid tube.

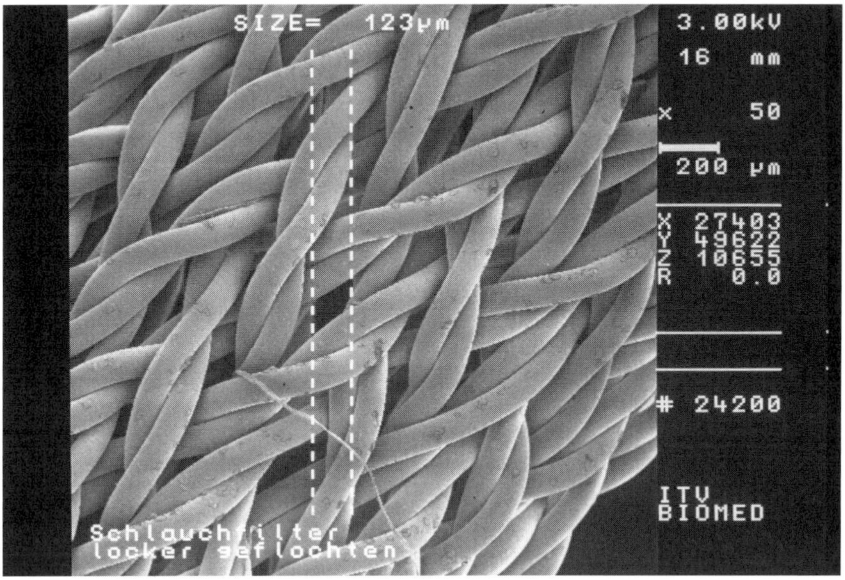

10.5 SEM photograph of a developed tube filter braid with twined monofilament fibers.

Figure 10.6 shows a rhombic structure with four equal sides of length *b*. If tractive forces F_Z are applied, the braid contracts, changing the braid angle and thus the pore size ΔU. Figure 10.7 shows the dependence of the pore size ΔU (in mm) on the braid angle of a tube filter ($d = 2.2$ mm, $b = 193$ µm). The particle separation barrier is principally defined by the diameter of the monofilament fiber and can be realized technically up to a fineness of 40 µm.

Assembly of filtration test equipment and production of filter tubes

A filtration test device consisting essentially of an inlet pump, a collection container and the filtration unit was assembled for wet filtration experiments. The sample to be filtered is pumped from the collecting container through a filtration circuit, which consists of the tube filter and a transparent acrylic pipe, and which demonstrates the filtration effect very effectively. The nominal diameters of the manufactured filters initially were 3, 4 and 5 mm.

The filtration unit is characterized by the fact that the pore size and thus the separation barrier can be modified by a factor of 3 by using clamps to alter the length of the tube filter in the piping module.

Determination of characteristic filtration data

A wet filtration system was used to attempt to separate different particles (e.g. coal

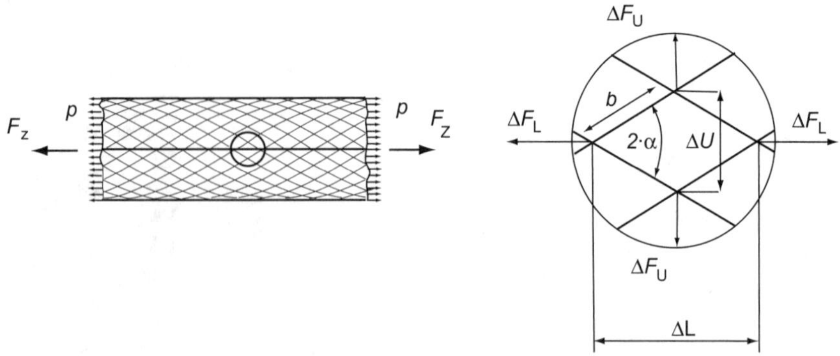

10.6 Geometrical analysis of a tube filter.

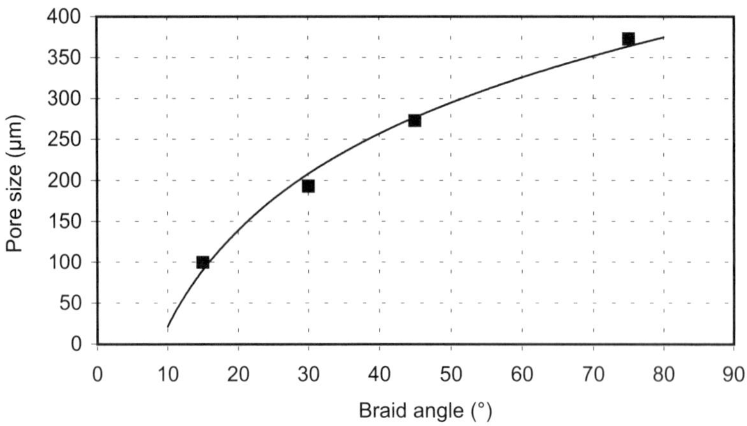

10.7 Pore size ΔU as a function of the braid angle α.

dust, polymer particles) with distinct particle sizes (15 to 355 μm). Figure 10.8 shows the sum distribution curve for a filtration trial with polymer particles. The characteristic curves for the collecting container and the filtrate are shown with stretched and relaxed tube filters. The external to internal filtration took place under a pressure difference of 10 to 50 Pa, a flow rate of 0.8 l min^{-1}, and a filter surface of approximately 0.04 m^2; the specific flow rate was 20 l min^{-1} m^{-2}).

Comparing the filtration characteristics demonstrates that, with the tube in a stretched condition, separation took place starting from a particle diameter of 30 μm. In a relaxed condition, separation begins at approximately 100 μm. Figure 10.9 shows the tube filter in stretched and relaxed conditions during filtration.

An additional process engineering advantage emerged during the filtration. Dynamic movements of the filter elements led to an effective separation of the surface layers formed and provided a cleaning function.

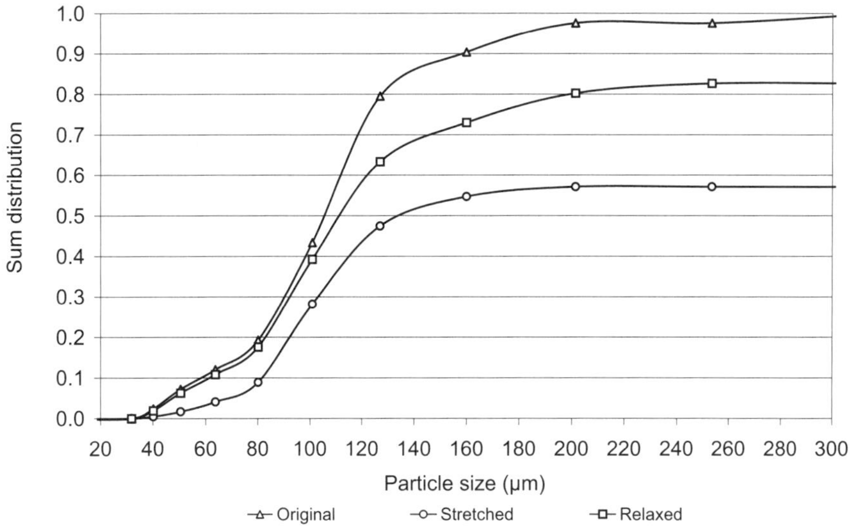

10.8 Sum distribution curve for a filtration experiment with polymer dust.

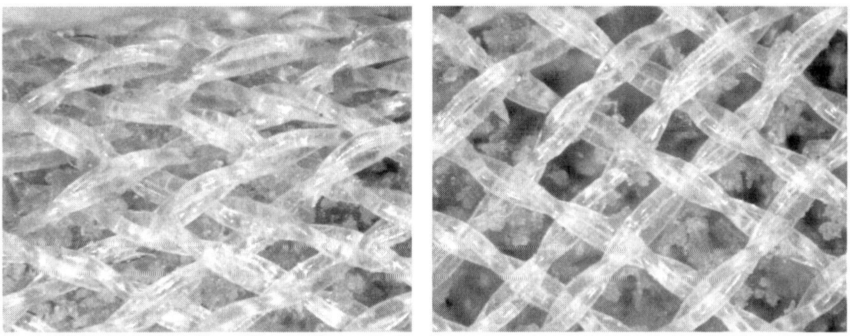

(a) (b)

10.9 Tube filter (a) stretched and (b) relaxed.

Outlook

Variable pore size and good cleaning behavior suggest that these adaptive tube filters can offer new micro-filtration methods in the fields of waste water, and food and chemical technology. Current work is aimed at producing industrial modular filter elements for these different applications.

10.4.2 Liquid transport without the use of pumps

Liquid can be transported in fiber-based systems without the use of pumps. New

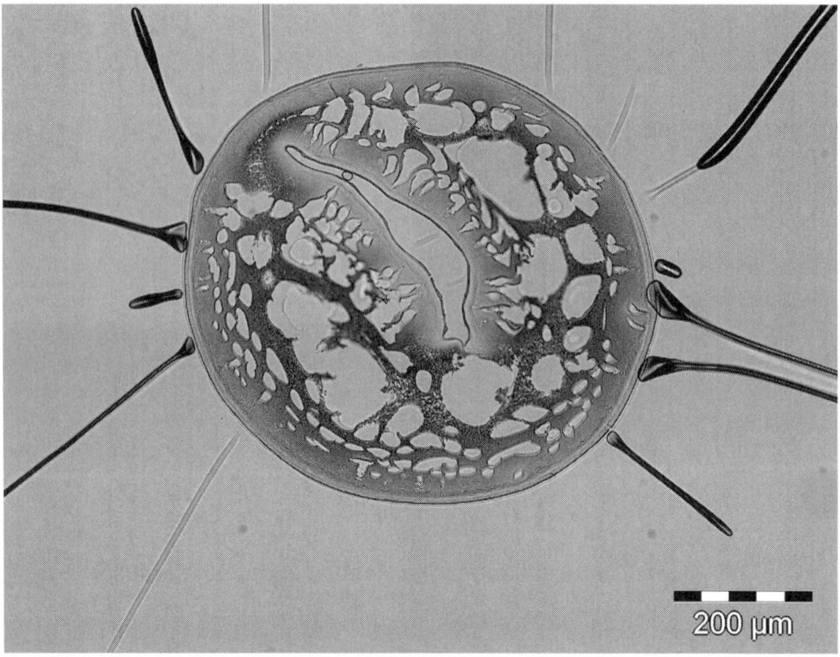

10.10 Cross-section of polymer hollow fiber.

technical textiles for long-distance transport of low-viscosity liquids are being developed in cooperation with Tübingen University, based on the model of water transport systems in plants (Fig. 10.10).

Trees and vines can transport water over great heights and distances without mechanical pumping systems and without requiring additional energy input. The energy necessary is supplied by the sun, suction being provided by transpiration (evaporation) at the end of the system – the pores, or stomata, in the leaves. In terms of utilizing this in biomimetically inspired textile materials, the following properties are of interest:

- the transport volume exclusively regulates itself according to need: only the water required for metabolism processes in the plant and transpiration is transported;
- transport security is maintained by preventing embolism.

Climbing vines can efficiently and securely transport water over several hundred meters via specialized water-carrying cells in the xylem, and serve as an interesting source of creative ideas for new technologies. There is a danger that gas bubbles might develop in the xylem because of high internal negative pressure, which would interrupt the flow and thus stop water transport (embolism). Various mechanisms contribute significantly to the prevention of embolism, based on the micro-morphology and biochemistry of the inner surfaces of the xylem cells and the membrane valves which connect them. To date, we have yet to develop technical solutions to the challenge of transporting liquid over long distances without mechanical pumping systems.

Possible applications of biomimetically inspired textiles include underground irrigation systems that do not require active pumping mechanisms and which provide more economical, tailored water release, textiles that transport large volumes of liquid for medical use, the clothing industry and fuel cells (carrying away the water formed at the fuel cell membrane).

10.5 Research activities in energy technology/ management

10.5.1 Transparent heat insulation for solar–thermal applications

Solar energy

Fossil energy sources are finite, and the development of efficient solar heating is an essential task of our time. Solar radiation at the earth's surface contains about 3% ultra violet (UV), 46% visible (UV–VIS) and 51% infra red (IR) radiation. The insolation power at a particular point on the earth's surface corresponds to its latitude. The annual insolation power for Munich shows a maximal value of 1.1 kWh m^{-2}, whilst in the Sahara it amounts to 2.2 kWh m^{-2}.

Thermal solar collectors change the sun's radiation power into usable heat. There are two different types of collector, concentrating high-temperature collectors (HT collectors), where the solar radiation is reflected by one or more hollow mirrors and collected in a receiver, and non-concentrating flat or low-temperature collectors (NT collectors), which collect solar radiation by means of an absorber layer or tubes. The absorbers, which consist of efficient heat-conducting metals or plastics, transfer the heat via a carrier medium (air, water, glycol water mixtures). Absorbers should have a high absorption rate and convert solar radiation as completely as possible into heat.

Materials used for the covers of solar collectors and translucent thermal insulation (TTI) on buildings have to have high translucence and, at the same time, high thermal insulation characteristics. Insulation glasses with excellent optical characteristics and insulating materials with fine capillaries or honeycombs arranged side by side are used as TTIs. They are most effective when the incidence of the sunlight is vertical.

To date, the materials used for absorbers and TTIs are plate-shaped, inflexible and rigid. They are also heavy and fragile due to the panes of glass, and so only suited for local use. A new, flexible transparent heat insulation material has been

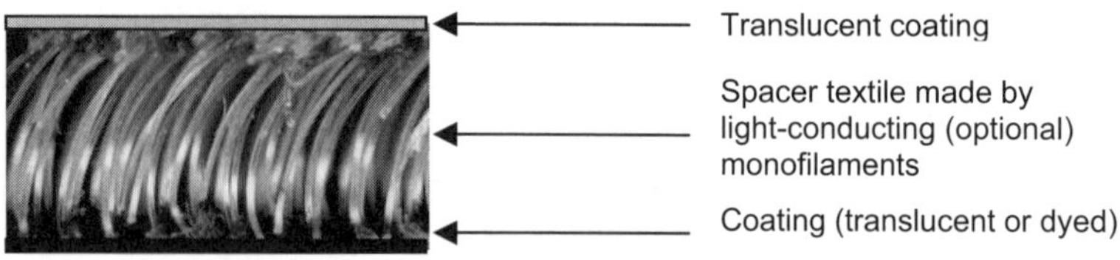

10.11 Construction of a spacer textile composite.

developed based on the solar–thermal functions of the pelt (skin and fur) of the polar bear.

Development of translucent thermal insulation material based on polar bear pelt

The possibilities for the development and production of flexible collectors were analyzed under a European Union research project.[5] Flexible materials for absorbers are already well known in principle, but they are unsuitable for TTI.[6] Polar bears have to survive in the arctic cold.[7] They have a black epidermis and opaque skin, and incident sunlight is transmitted by means of the yellowish white hair to the skin and transformed into heat.

At ITV Denkendorf, this principle was analyzed in detail. Solar power was transferred to textiles, which could be used in industrial solar–thermal applications (heating water and air) as flexible, translucent heat insulators. The product, based on coated spacer structures, can be manufactured on a large, industrial scale. Spacer textiles with translucent and/or dyed coatings showed particularly good performance. Figure 10.11 shows schematically the structure of a spacer textile with a double-sided coating.

The spacer textile is characterized by the following properties:

- highly light-resistant polymers,
- highly translucent and/or black pigmented silicone coating,
- translucence to incident light in the visible spectrum and impermeability to UV radiation,
- heat loss by convection strongly reduced,
- coating to prevent heat loss via long-wave (thermal) radiation,
- dirt-resistant coating, providing good translucence and high thermal efficiency.

The TTI textile shows some special advantages compared with other thermal insulation materials:

- relatively low weight,
- high mechanical stability (unbreakable, tearproof, elastic),
- high thermal stability (approximately up to 160 °C),
- flexibility, i.e. arched structures are feasible,
- deep-drawable within certain limits,

Table 10.1 Technical data of different translucent heat insulation materials

	Spacer textile	Hollow chamber panel	Comb structure
Thickness (mm)	5–60	6–16	20–60 (28–68[1])
Mass per unit area (kg m^{-2})	1.2–2.0	1.3–3.1	0.32–0.96 (20.32–20.96[1])
Light transmission (%)	80–≤95	77–82	84
Thermal transition coefficient *U*-value (W K^{-1} m^{-2})	2.2–3.0	2.6–3.6	1.3–2.2

[1]With double panes of glass.

- chemical resistance due to the silicone rubber coating,
- dirt resistance due to a special surface treatment (for self-purification, water is sufficient).

Table 10.1 shows the technical data for a translucent spacer textile coated on both sides (Fig. 10.12) and for commercially available TTI hollow chamber panels and structures, which are inserted in double panes of glass. The properties of the spacer textiles can be adjusted over a wide range by their construction and the coating conditions. Comparing the materials shows that the flexible spacer textiles, having

10.12 Translucent coated spacer textiles.

a low weight, high light transmission and low thermal transition coefficient (*U*-value), have advantages over the TTI materials used at present.

Potential applications

The company Solarenergie Stefanakis, one of the project partners, is using this new, flexible, textile-based construction for a new generation of solar collectors. Solarenergie Stefanakis has been developing and building spherical, hemispherical and parabolic thermal solar collectors for warm water production for many years. The spherical and hemispherical forms give the collectors the largest aperture areas possible, and the collectors do not have to be aligned according to the sun's position. The new materials have several advantages, including low prices, low assembly and maintenance costs and wind insensitivity. The spherical and hemispherical collectors are used as tank and tube absorbers, consisting of copper tubes or fabric hoses, and are suitable as continuous flow water heaters (Fig. 10.13).

In earlier models, the absorber was covered with a heat insulating acrylic glass cupola to prevent heat losses (particularly at night), providing TTI due to the included air layer and prevention of convection (Fig. 10.13). The new spacer textiles were tested as substitutes for the acrylic glass cupola (Fig. 10.14). A translucent spacer textile coated on one side was tested. The lighter, elastic and unbreakable spacer textiles show high thermal insulation effects. The textile can also be used as an absorber with integrated thermal insulation (Fig. 10.15), if one side is translucent and the other side is coated with a black absorbing layer. These materials are not only useful for collectors, but also have applications in the construction industry with a modified structure, as front elements and for roof constructions that allow an architecturally free design.

10.5.2 Adaptive breathable membranes for climate regulation

In cooperation with Tübingen University, we have analyzed the possibility of using the mechanism that controls water evaporation in plants to develop new materials with the potential to optimize liquid transport via self-regulating micro-pores, e.g. for clothing, upholstery, and wound dressing. Detailed anatomical studies of the stomata (plant pores) were carried out using SEM and 3-D microscopy, paying particular attention to the micro-structures responsible for optimizing the transport of liquids. Computer simulations of transpiration at the stomata were then developed, and revealed the effect of the different features on liquid transport. This enabled us to identify the structures that are of functional importance for evaporation-induced liquid transport, resulting in two basic conceptions:

10.13 Hemispheric collector with tube absorber from the company Solarenergie Stefanakis.

- Lamination to create humidity-sensitive coating systems using a combination of layers of different materials. According to the basic idea, two porous materials must be stacked in such a way that the pores of one layer are covered by the corresponding pores on the other. These materials are coated with swelling media which expand in the presence of humidity and thus promote evaporation through the pores.
- Combination of two fiber types, one that is very sensitive to humidity and lengthens in the presence of humidity, and one that is not sensitive to humidity.

Samples were produced and the adaptive function regarding permeability to air and steam was proved. Impermeability to water penetrating from the outside was achieved by including an external water-repellent layer.

10.14 Deep-draw (formed with heat and pressure) spacer textile as collector cover.

10.15 Spacer textile with translucent and black coating.

10.6 Research activities in the field of lightweight construction

The search for lightweight construction materials led us to look at natural, weight-optimized structures such as bones and plant stems. In nature, the fibers are orientated optimally according to the main direction of stress, and the number of fibers is just sufficient to provide the necessary strength. Research activities were aimed at developing cheap textile technologies that made it possible to include reinforcing fibers in response to the expected distribution of forces, for instance for the production of tubes that must withstand bending, buckling or torsional stress. Findings from the Freiburg University Plant Biomechanics Group concerning the structure of stalks were used to produce profiles of composites with optimized weight-related bending stiffness/strength and high dynamic load capacity and damping behavior.[4]

Plant stems can be understood as composite, fibrous materials that are composed of various materials with different mechanical properties. The bionic potential of hollow, cylindrical plant stems, including those of the horsetail (*Equisetum hyemale*) and giant reed (*Arundo donax*), were investigated.

The stem of the horsetail plant (Fig. 10.16) is composed of external and internal pressure cylinders and connecting spacers. There are variations at different hierarchical layers; for instance, an optimal fiber layout includes a higher fiber volume fraction and thicker cell walls in the outer parts of the cross-section. This structure represents a lightweight construction with high, specific bending stiffness and buckling stability using the minimal amount of materials. The horsetail stem demon-strates the well-known fact that, similar to double T-supports, the main mass of the supports is stored in the pull and pressure belts, respectively, at a distance as large as possible from the neutral bending fiber. Based on the different biological models, a general model for lightweight tubular structures with high bending stiffness was defined, and this formed the basis for studying potential industrial applications.

The giant reed plant (*Arundo donax*), as well as optimizing fiber weight, shows excellent damping behavior towards mechanical vibrations along with favorable breaking behavior (with several pre-failure events). This is because there is a gradual transition in stiffness between fibers and basic matrix.

A braid pultrusion technology, specially adapted and suitable for large-batch applications, was developed and installed at ITV Denkendorf to utilize this ultra-light sandwich structure. The yarns stored on the bobbin creel and on the braiding carriers are led through a matrix-impregnation bath and a heated pultrusion tool. The matrix of impregnated fibers is cured while being led through the tool. Spiral-shaped fibers can be produced in the profile by integrating a braiding machine into the pultrusion line, and the braided fibers impart desired properties such as torsional stiffness, high vibration damping and favorable structural integrity. Using this braid pultrusion technology, sandwich structures for fiber composite

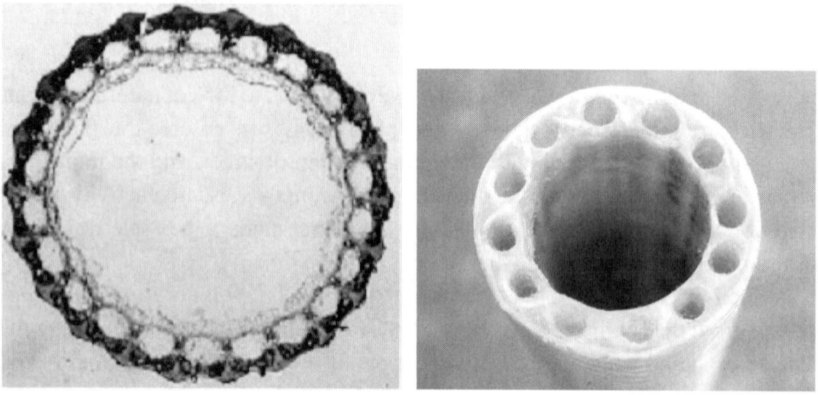

(a) (b)

10.16 (a) Horsetail plant (*Equisetum hyemale*) (b) technical plant stems (both in cross-section).

profiles with tubular cross-sections, so-called 'Technical plant stems', were produced.

There is a wide range of technical applications for these innovative high-performance fiber composite profiles, covering all applications which require tubular fiber composite structures, including aerospace technology, vehicle construction, building technology, instrument building, medical engineering (prosthetics) and sports equipment.

10.7 Future trends

Nature provides technicians with a very broad field of highly interesting models. Continuous development of instruments, detectors and sensors will enable us to understand the functions and mechanisms of the natural world, and translate this information into useful technologies. Nanoscience, in particular, should lead to significant advances.

10.8 Conclusions

Bionic research in fiber-based materials has great potential for innovation and optimization. At ITV Denkendorf, bionics is a strategic pillar of future research aimed at developing new materials and opening up new markets.

Thanks to the support of public funding, it has become possible to conduct the necessary basic research and take important steps in product development. Applied research activities with potential partners from industry have made it possible to bring the first bionic products to market. This collaboration will continue.

10.9 Sources of further information and advice

www.itv-denkendorf.de
www.biokon.de
www.kompetenznetz-biomimetik.de

10.10 Acknowledgements

We would like to thank the Ministry of Science, Research and the Arts Baden-Wurttemberg for funding the Network for Biomimetics Baden-Wurttemberg, which laid the foundation of interdisciplinary bionic research in Baden-Wurttemberg.

Also thanks to the Federal Ministry of Education and Research (BMBF) for sponsoring important feasibility studies that allowed us to pinpoint the potential of bionic ideas.

Moreover, we would like to express our thanks to 'Deutsche Bundesstiftung Umwelt' and the European Union for funding projects to realize ideas which have already resulted in prototypes and industrial products.

Finally, we would like to thank 'Forschungskuratorium Textil e.V.' for financially supporting the research project (AIF-Nr. 13573N) 'Influence of structure and orientation of staple fibers as well as filaments on extremely self-cleaning Lotus-surfaces' that resulted from the funds from the Federal Ministry of Economics and Technology (BMWi) by means of a grant from Arbeitsgemeinschaft industrieller Forschungsvereinigungen 'Otto von Guericke' e.V. (AIF).

10.11 References

1 Arnim, V. v., Scherrieble, A., Reichardt, S., Stegmaier, T. and Planck, H., 'Lotus Effekt auf Textilien – Grundlagen und Anwendung' Vortrag auf dem Denkendorfer Nano Forum 'Selbstreinigende Textilien – Das Gütesiegel für Textilien mit dem Lotus-Effekt', SI-Centrum Stuttgart, 6 April 2006.

2 Vogel, S., *Von Grashalmen und Hochhäusern – Mechanische Schöpfungen in Natur und Technik*, Wiley-VCH Verlag, 2000.

3 Steinmann, P., http://psteinmann.net/me.html (2004).

4 Lehmann-Maschinenbau GmbH, Dr. Bergemann, TU Chemnitz-Zwickau: Abschlussbericht 'Fest/Flüssig-Trennung durch kontrahierende Schlauchfiltermembranen', gefördert vom Freistaat Sachsen (PT 4301), 1999.

5 Institut für Textil- und Verfahrenstechnik (ITV) Denkendorf Development of flexible and extremely economic sun collectors (FLEXCOLL), Final report of the SME/CRAFT European Research Project (CRAFT-1999-71816), 2005.

6 Stegmaier, T., Abele, H., 'Textilien für Photovoltaik und solare Anwendungen', Lecture on High-Tex 2000, Stuttgart (Germany), 7/8 Juli 2000.

7 Nachtigall, W., *Bionik-Grundlagen und Beispiele für Ingenieure und Naturwissenschaftler*, Springer-Verlag Berlin Heidelberg, 1998.

Index